Formula for the future: nutrition or pathology?

Formula for the future: nutrition or pathology?

Elevating performance and health in pigs and poultry

edited by:

J.A. Taylor-Pickard

Z. Stevenson

K. Glebocka

Wageningen Academic
P u b l i s h e r s

ISBN: 978-90-8686-088-3

First published, 2008

**Wageningen Academic Publishers
The Netherlands, 2008**

Contents

Regulating gut function and immunity

Denise Kelly, Imke Elisabeth Mulder and Bettina Schmidt
Rowett Institute of Nutrition & Health, University of Aberdeen, Greenburn Road,
Bucksburn, Aberdeen AB21 9SB, United Kingdom

1. Introduction

The mammalian gastrointestinal (GI) tract contains a diverse population of micro-organisms, collectively known as the gut microbiota, the development of which begins at birth following the acquisition of both maternal and environmental microbes. The gut microbiota quickly expands into a complex and dynamic ecosystem that fulfils important nutritional, immunological and health-promoting functions in the host. These functions reflect the long co-evolutionary existence between microbiota and host gut and are a consequence of symbiotic host–microbe interactions, that influence a diverse range of physiological processes including lipid metabolism of the host (Backhed *et al.*, 2004; Ley *et al.*, 2006), immune development and homeostasis, inflammation, repair and angiogenesis (Rakoff-Nahoum *et al.*, 2004; Kelly *et al.*, 2004; Stappenbeck *et al.*, 2005). In this short review the contribution of the gut microbiota to gut function and mucosal immunity will be discussed.

2. Effects of the commensal microbiota on the host gut maturation

The profound impact of the microbiota on gut function is best illustrated by comparison of the germ-free with the conventionally-colonised gut. Germ-free animals have a poorly developed mucosal immune system, with reduced numbers of lymphocytes, plasma cells and mononuclear cells (Wostmann *et al.*, 1996). Histologically, the villus-crypt architecture is also immature and changes dramatically upon microbial colonisation indicating that the commensal microbiota influences both processes of epithelial cell turnover rate and cellular differentiation in the gut epithelium (Umesaki *et al.*, 1993; Alam *et al.*, 1994). Indeed, several studies suggest that the microbiota can influence the expansion of secretory cells such as mucus-secreting goblet cells (Sharma and Schumacher, 1995). These cells are less abundant in the gut epithelium of germ-free rats and any mucus generated remains undegraded in an enlarged caecum. The germ-free gut remains morphologically underdeveloped until the introduction of intestinal contents or faeces from conventional animals. This process, referred to as conventionalisation, stimulates the normal physiological and immunological characteristics of the gut, further strengthening the evidence of the functional importance of the normal commensal microbiota in gut development (Matsumoto *et al.*, 1992; Umesaki *et al.*, 1993). In addition to the overall importance of commensal microbe exposure to gut development and function is the notion of a 'developmental window', when exposure to commensal microbes, particularly during early life, can have a long-lasting impact on health status (Thompson *et al.*, 2008). From a practical standpoint this is an important concept as it places greater emphasis on the quality of nutrition and the rearing environment in early life as a means of optimising life-time performance.

3. Host-microbe interaction: role of epithelial cells

The intestinal epithelial lining provides an important barrier between the host and the environment and protects the body against invasion and systemic dissemination of both pathogens and commensal micro-organisms. This surface also represents an important point of contact with gut microbes and supports the regulated sampling of bacterial antigens and signalling events, both of which are critical for immune development. These antigen-independent signals are mediated through cell surface receptors and specific microbial ligands. They are driven by interactions between pattern-recognition receptors (PRRs) and specific microbial products, referred to as pathogen-associated molecular patterns (PAMPS) (Butler *et al.* 2000). PAMPS are structures commonly found on both commensal and pathogenic bacteria and include lipopolysaccharide (LPS) and flagellin. These structures interact with PRRs, also referred to as toll-like receptors (TLRs) (Creagh and O'Neill, 2006). TLR4 and TLR5 interact with LPS and flagellin respectively and function to alert the immune system to the presence of bacteria. One key question for mucosal immunologists has been to explain how the gut of mammals can harbour such abundant levels of microbial LPS without mounting strong immune responses, which if excessive can be damaging. Recent published evidence is beginning to provide some understanding. Bates *et al.* (2007) have shown that alkaline phosphatase, an enzyme expressed by intestinal epithelial cells, localised to the gut lumen, and induced by microbial colonisation can detoxify LPS and hence plays an important role in maintaining immune homeostasis. In addition, to their function in signalling the presence of bacteria, epithelial cells also play an important role in controlling the composition of the microbial community. This is achieved through the synthesis and secretion of anti-microbial peptides (Ryu *et al.*, 2008). Understanding how to manipulate anti-microbial peptide secretion in the host gut is considered a new and exciting route to promoting 'natural' disease resistance by preventing pathogenic colonisation and promoting the establishment of beneficial gut bacteria.

4. Host-microbe interaction: role of mucosal dendritic cells and antigen presentation

The mucosal immune system in the GI tract consists of several important lymphoid networks, which play an important role in bacterial/immune cell interactions. These include Peyer's patches (PP), lamina propria (LP) and mesenteric lymph nodes (MLN). A wide variety of antigens are present in the gut, both on the mucosal surface and in the lumen, and PPs are involved in directly sampling antigens from the lumen. Antigens are taken up across the follicle-associated epithelium (FAE) of the PPs and transported through specialised M cells (microfold cells) to dendritic cells (DCs). DCs are professional antigen presenting cells that migrate to the MLNs and interact with naïve T cells which results in T cell activation, maturation and migration. In addition to PP, microbial antigens are also sampled directly by dendritic cells lying beneath the intestinal epithelium. This unique DC subset functions by extending dendrites through the epithelium and requires manipulation of tight-cell junctions, allowing direct antigen sampling (Rescigno *et al.*, 2001; Macpherson and Uhr, 2004). Following antigen acquisition, mucosal DCs transfer through afferent lymphatics to the MLN, where they can present antigen to T cells (Huang *et al.*, 2000).

In addition to the functions of the PP and 'interdigitating' epithelial DCs, the LP of the pig intestine shows a high level of organisation. Antigen-presenting cells expressing MHC II are present in large numbers in the LP of adult pigs (Wilson *et al.*, 1996) and they have been characterised as functional, immature, dendritic cells (Haverson *et al.*, 2000). Cells of the dendritic lineage are also present in large numbers within the villi and co-localise with T cells expressing the CD4 co-receptor. This cluster of MHC class II$^+$ DCs, endothelium and CD4$^+$ T cells is also thought to function as a complex antigen-presenting zone in the pig intestine (Bailey *et al.*, 2005).

5. Mucosal immune development in the pig: innate and adaptive immunity

The general principles of bacterial signalling and antigen presentation highlighted in the above sections apply universally to the development of the mammalian immune system. Hence microbial colonisation is a key driver for the development of the pig mucosal immune system and involves microbial induced TLR receptor signalling, a key component of the innate immune defence system and immune-education through antigen-specific presentation and T/B cell expansion and maturation, which are prerequisite events driving the development of the adaptive immune system. This complex system of antigen recognition, sampling and presentation is underdeveloped in the newborn pig. Development of this system occurs with age, and is strongly influenced by environmental factors including microbial colonisation, exposure to specific antigens and nutrition. In the neonatal pig, mucosal development takes place in four phases between birth and 6 weeks of age and involves expansion and maturation of B cells, T cells and antigen presenting cells (Bailey *et al.*, 2005). By week seven the architecture of the intestine is comparable to that of a mature animal. Clearly, protection against pathogenic challenge requires a rapid activation of both arms of the immune system to ensure pathogen clearance, whereas maintaining a diverse commensal microbiota necessitates immune tolerance. Immunological tolerance is achieved through several mechanisms, one of which involves the induction and presence of T regulatory cells (Tregs). Recently, a commensal bacterium, *Bifidobacterium infantis* 35624 has been shown to be able to induce Tregs (O'Mahony *et al.*, 2008) and it has been suggested that this micro-organism contributes to host immune homeostasis.

6. Development of the commensal microbiota

The microbiota of the GI tract is a relatively stable community that has successfully established over the period between birth and adulthood. *In utero*, the GI tract of the pig is sterile (Kelly and King, 2001). From the moment the piglet is born however, a wide variety of microbes originating from the vagina, faeces and skin of the sow, as well as the environment, are ingested and colonise the GI tract with varying degree of success (Mackie *et al.*, 1999). Thereafter, the establishment, competition and succession of microbial populations within the gut is influenced by a complex interplay between environmental, nutritional, microbial and host factors that eventually dictate the composition and stability of the microbial population inhabiting specific regions of the GI tract. Significant progress in our knowledge of microbial diversity in the mammalian gut has been achieved through molecular techniques such as phylogenetic analysis of 16S ribosomal DNA (rDNA) sequences (Leser *et al.*, 2002; Eckburg *et al.*, 2005; Ley *et al.*, 2008). Leser *et al.* (2002) have described the gut microbiota of the pig based on approximately 4,000 16S genes and found that the majority of identified phylotypes (around 81%) belonged to low-G+C gram-positive bacteria (*Streptococcus, Staphylococcus, Bacillus, Clostridium,*

Lactobacillus), and 11.2% belonged to *Bacteroides* and *Prevotella* groups. This data also revealed that 83% of the sequences were unknown. Similarly, direct comparisons of the phylogenetic analysis of 16S rDNA of the microbial community with data generated by culture-based methods indicates that 60 to 80% of the micro-organisms in the total microbiota have not been cultivated previously (Suau *et al.*, 1999; Salzman *et al.*, 2002; Leser *et al.*, 2002; Eckburg *et al.*, 2005; Ley *et al.*, 2008). The contribution of these uncultured micro-organisms to the maintenance of immune tolerance and mucosal defence of the GI tract is currently unknown, but is likely to be important.

7. Gut microbiota and mucosal localisation

The normal adult microbiota increases numerically from the upper to the lower GI tract (Zoetendal *et al.*, 1998). The stomach and small intestine contain low numbers of bacteria as few bacterial groups can overcome the acidic conditions and rapid flow of digesta in this region of the GI tract. The jejunal and ileal luminal microbiota consists of mostly facultative anaerobes, such as *Lactobacilli*, *Streptococci*, *Enterococci*, gamma-proteobacteria, and *Bacteroides* species (Hayashi *et al.*, 2005). The caecum and colon are the main sites for bacterial colonisation in the gut, and the microbiota in this region forms an increasingly complex ecosystem. High substrate availability, slow passage rate and higher pH create a favourable niche for the colonisation of a diverse population of bacteria which number up to 10^{12} colony-forming units/g of digesta (Pryde *et al.*, 1999). The majority of culturable bacteria include gram-positives such as *Streptococci*, *Lactobacilli*, *Peptostreptococci* and *Clostridia*, and gram-negative bacteria affiliated with *Prevotella* and *Bacteroides*.

In terms of the functionality of the GI microbiota, much of the current knowledge has been based on the analysis of faecal samples. However, recent data suggests that the mucosa-adherent (mucosa-associated) and luminal microbial populations (represented by faecal analysis) are in fact distinct populations and hence may fulfil different roles within the gut ecosystem (Pryde *et al.*, 1999; Zoetendal *et al.*, 2002; Eckburg *et al.*, 2005). This may be a potentially important consideration in identifying bacteria which influence gut function and immunity; it could be argued that bacteria that have evolved to gain close access to gut surfaces exert greater influence on physiology and functionality of the gut than luminal micro-organisms that colonise gut contents and dietary substrates. To further complicate matters, significant differences in adherent microbial populations also exist between individuals (Green *et al.*, 2006) indicating the potential importance of the host genotype in determining the composition of mucosa-associated micro-organisms. This latter point supports the notion of functional redundancy within the microbial community and implies that different bacteria may provide similar/compensatory immune functions.

8. Host-microbe interactions and disease susceptibility

Although the microbiota is an important factor in promoting the development and maintenance of the immune system in early life, morbidity and mortality resulting from infections during the immediate post-weaning period are of frequent occurrence and represent a major financial loss to the pig industry. Factors including removal from the sow and littermates at weaning, reduced feed intake, introduction of new dietary antigens from the weaning diet, removal of maternal immune protection via milk, disruption of the microbiota and the pathogen-load of the weaning environment all contribute to the increased disease risk of the piglet during the weaning process. Post-weaning diarrhoea is mainly caused by enterotoxigenic *Escherichia coli* K88 (Moon and Bunn, 1993; Osek, 1999). Porcine circovirus 2 is the causative pathogen associated with post-weaning multisystemic wasting syndrome (PMWS). Other diseases that have a financial impact on pig health during this time include Rotavirus, *Salmonella*, Swine Dysentery and Haemophilus. Even beyond the weaning period, non-specific colitis, a disease with striking similarities to human inflammatory bowel disease, is a serious problem in growing and finishing pigs, and negatively impairs animal performance.

Current methods of disease control rely on the sub-therapeutic and clinical use of antibiotics (Dritz *et al.*, 2002; Gaskins *et al.*, 2002). Inclusion of these substances in animal feeds as growth promoters has long been known to improve growth rates and feed conversion efficiency, and to protect against pathogenic bacteria. A European ban on the use of antibiotic growth promoters however, has been recently implemented due to both the prevalence of antibiotic-resistant bacteria and environmental concerns (Anadon, 2006). This ban has put greater pressure on the industry to find alternative methods and strategies to prevent/treat infectious diseases. With the accumulating evidence for a generic role of the commensal microbiota in driving and promoting immune defense mechanisms and the identification of microbe-specific effects, the possibility of either engineering the microbial community and/or introducing new bacterial strains to achieve prophylactic/therapeutic endpoints is now being tested widely within both the human and animal health sectors.

9. Disease susceptibility and the hygiene hypothesis

Epidemiological evidence exists to support the view that environmental and microbial exposure can profoundly influence gut health. This has given rise to the so-called 'Hygiene Hypothesis' which postulates that allergic and autoimmune diseases, both of which are increasing in Westernised countries, are a consequence of reduced childhood infections and the high-hygiene status of the western lifestyle (Strachan, 1989; Wills-Karp *et al.*, 2001). Animal studies give support to this hypothesis as susceptibility to autoimmune diseases increases when animals are bred under germ-free conditions; disease is prevented when the animals are exposed to bacteria (Bach, 2005). The relevance of the hygiene hypothesis to the normal intensive rearing conditions used in animal husbandry is not clear, but it is likely that rearing environment and ultimately, environmental exposure to bacteria and other micro-organisms has a profound influence on both immunity and disease resilience, analogous to the human situation (Thompson *et al.*, 2008). The recent move towards organic farming practices, in addition to meeting with current social and welfare requirements, may also address the need for adequate microbial exposure to promote immunological function.

10. Identifying functionally important gut bacteria with prophylactic and therapeutic properties

The effects of the commensal microbiota, and indeed probiotics, on the intestinal immune development are strain-specific. To illustrate this phenomenon, germ-free animals mono-colonised with segmented filamentous bacteria (SFB) showed an improved immune development in the small intestine, but not in the remaining gut segments. Conversely, immunological development in the large intestine was restored by colonisation with Clostridia, a common bacterial group of the large intestine. In the presence of both SFB and Clostridia, almost the same level of development of the gut immune system occurred as observed in the conventionalisation process (Umesaki *et al.*, 1999). This study demonstrates not only, that different commensal species exert their immune-stimulating effects in different locations within the gut, but also the complexity by which the commensal microbiota interacts to modulate the host's immune system.

The idea that specific gut bacteria exert health-promoting effects on gut function and immunity has fuelled the huge expansion in the consumption/administration of probiotics and prebiotics products in both the human and animal health sectors (Hord 2008; Gibson, 2008; Lalles *et al.*, 2007).

11. Prebiotics and probiotics: natural alternatives to antibiotics

The application of dietary ingredients to target the gut microbiota, in order to improve intestinal function and a healthy gastrointestinal tract forms the underlying rationale of the use of prebiotics. Prebiotics are food additives that promote colonisation of so-called 'good' bacteria with health benefits for the host (Tako *et al.*, 2008). Commonly investigated prebiotics include inulin and inulin-type fructans (Tako *et al.*, 2008). Inulin improves iron and calcium absorption, and is fermented in the colon to produce short chain fatty acids (SCFA) such as butyrate, thereby stimulating the growth of health-promoting bacteria such as Bifidobacteria and Lactobacilli (Tako *et al.*, 2008). Butyrate also has direct effects on colonic health, by regulating processes associated with proliferation, differentiation and apoptosis of epithelial cells (Tako *et al.*, 2008).

Probiotics are defined as 'Live micro-organisms which when administered in adequate amounts confer a health benefit on the host' (FAO/WHO, 2001). Most probiotic products contain *Lactobacillus* and/or *Bifidobacterium* and a large number of trials have reported either successful, variable or negative results. These types of experiments illustrate our incomplete knowledge of the factors that contribute to successful bacterial colonisation, particularly in the context of an established microbial ecosystem, but also raise questions as to whether the bacterial strains currently used in probiotic products represent the best strain selection to promote immune function and disease resistance.

Probiotic use in pigs has been reported to improve growth performance, decrease the incidence of diarrhoea and decrease morbidity/mortality rates (Lalles *et al.*, 2007). The probiotic species most commonly investigated in pigs are Lactobacilli, Enterococci and yeast (Takahashi *et al.*, 2007). The lactic-acid producer *Pediococcus acidilactici* and the live yeast *Saccharomyces cerevisiae* ssp. exert beneficial effects on gut structure and function (Di Giancamillo *et al.*, 2008; Baum *et al.*, 2002; Bontempo *et al.*, 2006). Probiotics are thought to prevent pathogen colonisation through competition for biological niches determined by oxygen concentration, pH or nutrient requirements. Probiotic bacteria can also produce antibacterial compounds including hydrogen peroxide, organic acids and bacteriocins. For instance, the Abp 118 bacteriocin produced by *Lactobacillus salivarius* has anti-listerial effects in mice (Corr *et al.*, 2007). Finally, as mentioned before, these bacteria also potentially modulate the systemic immune response through the mucosal immune system (Corth *et al.*, 2007).

12. Conclusions

Clearly important functional interactions between the commensal microbiota and the mucosal immune system exist and have developed over a long period of co-evolution. These interactions are being actively researched and dissected and are providing greater insight into the complex ways in which bacteria regulate host immune development and mucosal immune homeostasis. Such studies are also providing knowledge of potential bacterial candidates that trigger beneficial health effects of greater physiological significance than the current generation of probiotic and prebiotic products. Many of these methods are reliant on new molecular strategies; for example, 16S rRNA and metagenomic profiling of microbial communities in the healthy and diseased gut, in combination with analyses of microbial genomes. These approaches will hopefully identify new drug targets for the manipulation of microbial community structure and function (Zaneveld *et al.*, 2008). Ultimately, such manipulations should provide novel and effective ways of managing gut health during both early and adult life in both animals and humans.

Acknowledgements

The authors acknowledge the support of the Rural and Environment Research and Analysis Directorate (RERAD).

References

Alam, M., Midtvedt, T. and Uribe, A. (1994). Differential Cell Kinetics in the Ileum and Colon of Germfree Rats. *Scandinavian Journal of Gastroenterology* **29**: 445-451.

Anadon, A. (2006). The EU ban of antibiotics as feed additives (2006). Alternatives and consumer safety. *Journal of Veterinary Pharmacology and Therapy* **29**: 41-44.

Bach, J.F. (2005). Six questions about the hygiene hypothesis. *Cellular Immunology* **233**: 158-161.

Backhed, F., Ding, H., Wang, T., Hooper, L.V. and Koh, G.Y. (2004). The gut microbiota as an environmental factor that regulates fat storage. *Proceedings of the National Academy of Sciences USA* **101**: 15718-15723.

Bailey, M., Haverson, K., Inman, C., Harris, C., Jones P., Corfield, G., Miller, B. and Stokes, C. (2005). The influence of environment on development of the mucosal immune system. *Veterinary Immunology and Immunopathology* **108**: 189-198.

Bates, J.M., Akerlund, J., Mittge, E. and Guillemin, K. (2007). Intestinal alkaline phosphatase detoxifies lipopolysaccharide and prevents inflammation in Zebrafish in response to the gut microbiota. *Cell Host & Microbe* **2**: 371-382.

Baum, B., Liebler-Tenorio, E.M., Enss, M.L., Pohlenz, J.F. and Breves, G. (2002). Saccharomyces boulardii and *Bacillus cereus* var. Toyoi influence the morphology and the mucins of the intestine of pigs. *Zeitschrift für Gastroenterologie* **40**: 277-284.

Bontempo, V., Di Giancamillo, A., Savoini, G., Dell'Orto, V. and Domeneghini, C. (2006). Live yeast dietary supplementation acts upon intestinal morpho-functional aspects and growth in weanling piglets. *Animal Feed Science and Technology* **129**: 224-236.

Butler, J.E., Sun, J., Weber, P., Navarro, P. and Francis, D. (2000). Antibody repertoire development in fetal and newborn piglets, III. Colonization of the gastrointestinal tract selectively diversifies the preimmune repertoire in mucosal lymphoid tissues. *Immunology* **100**: 119-130.

Creagh, E.M. and O'Neill, L.A. (2006). TLRs, NLRs and RLRs: A trinity of pathogen sensors that co-operate in innate immunity. *Trends in Immunology* **27**: 352-357.

Corr, S.C., Li Y, Riedel, C.U., O'Toole, P.W., Hill, C. and Gahan, C.G. (2007). Bacteriocin production as a mechanism for the antiinfective activity of *Lactobacillus salivarius* UCC118. *Proceedings of the National Academy of Sciences USA* **104**: 7617-7621.

Corth, B., Gaskins, H.R. and Mercenier, A. (2007). Cross-talk between probiotic bacteria and the host immune system. *Journal of Nutrition* **137**: 781S-790S.

Di Giancamillo, Vitari, F., Savoini, G., Bontempo, V., Bersani, C., Dell'Orto, V. and Domeneghini, C. (2008). Effects of orally administered probiotic *Pediococcus acidilactici* on the small and large intestine of weaning piglets. A qualitative and quantitative micro-anatomical study *Histology and Histopathology* **23**: 651-664.

Dritz, S.S., Tokach, M.D., Goodband, R.D. and Nelssen, J.L. (2002). Effects of administration of antimicrobials in feed on growth rate and feed efficiency of pigs in multisite production systems. *Journal of the American Veterinary Medical Association* **220**: 1690-1695.

Eckburg, P.B., Bik, E.M., Bernstein, C.N., Purdom, E., Dethlefsen, L., Sargent, M., Gill, S.R., Nelson, K.E. and Relman, D.A. (2005). Diversity of the Human Intestinal Microbial Flora. *Science* **308**: 1635-1638.

FAO/WHO. (2001). Health and nutritional properties of probiotics in food including powder milk with live lactic acid bacteria. Report of a Joint FAO/WHO Expert Consultation on Evaluation of Health and Nutritional Properties of Probiotics in Food Including Powder Milk with Live Lactic Acid Bacteria. Available at: http://www.mesanders.com/docs/probio_report.pdf.

Gaskins, H.R., Collier, C.T. and Anderson, D.B. (2002). Antibiotics as growth promotants: mode of action. *Animal Biotechnology* **13**: 29-42.

Gibson, G.R. (2008). Prebiotics as gut microflora management tools *Journal of Clinical Gastroenterology* **42:** S75-79.

Green, G.L., Brostoff, J., Hudspith, B., Michael, M., Mylonaki, M., Payment, N., Staines, N., Sanderson, J., Rampton, D.S. and Bruce, K.D. (2006). Molecular characterization of the bacteria adherent to human colorectal mucosa. *Journal of Applied Microbiology* **100**: 460-469.

Haverson, K., Singha, S., Stokes, C.R., and Bailey, M. (2000). Professional and non-professional antigen-presenting cells in the porcine small intestine. *Immunology* **101**: 492-500.

Hayashi, H., Takahashi, R., Nishi, T., Sakamoto, M. and Benno, Y. (2005). Molecular analysis of jejunal, ileal, caecal and recto-sigmoidal human colonic microbiota using 16S rRNA gene libraries and terminal restriction fragment length polymorphism. *Journal of Medical Microbiology* **54**: 1093-1101.

Hord, N.G. (2008). Eukaryotic-microbiota crosstalk:potential mechanisms for health benefits of prebiotics and probiotics. *Annual Review of Nutrition* **28**: 215-231.

Huang, F.P., Platt, N., Wykes, M., Major, J.R., Powell, T.J., Jenkins, C.D. and MacPherson, G.G. (2000). A discrete subpopulation of dendritic cells transports apoptotic intestinal epithelial cells to T cell areas of mesenteric lymph nodes. *The Journal of Experimental Medicine* **191**: 435-444.

Kelly, D. and King, T.P. (2001). Digestive physiology and development in pigs. In: *The Weaner Pig. Nutrition and Management.* M.A. Varley and J. Wiseman, Eds. CABI Publishing, U.K.

Kelly, D., Campbell, J.I., King, T.P., Grant, G., Jansson, E.A., Coutts, A.G.P., Pettersson, S. and Conway, S. (2004). Commensal anaerobic gut bacteria attenuate inflammation by regulating nuclear-cytoplasmic shuttling of PPAR-[gamma] and RelA. *Nature Immunology* **5**: 104-112.

Klaasen, H.L.B.M., Van der Heijden, P.J., Poelma, F.G.J., Koopman, J.P., Van den Brink, M.E., Bakker, M.H., Eling, W.M. and Beynen, A.C. (1993). Apathogenic, intestinal, segmented, filamentous bacteria stimulate the mucosal immune system of mice. *Infection and Immunity* **61**: 303-306.

Lalles, J.P., Bosi, P., Smidt, H. and Stokes, C.R. (2007). Nutritional management of gut health in pigs around weaning. *Proceedings of the Nutrition Society* **66**: 260-268.

Leser, T.D., Amenuvor, J.Z., Jensen, T.K., Lindecrona, R.H., Boye, M. and Moller, K. (2002). Culture-independent analysis of gut bacteria: the pig gastrointestinal tract microbiota revisited. *Applied and Environmental Microbiology* **68**: 673-690.

Ley, R.E., Turnbaugh, P.J., Klein, S. and Gordon, J.I. (2006). Microbial ecology: Human gut microbes associated with obesity. *Nature* **444**: 1022-1023.

Mackie, R.I., Sghir, A. and Gaskins, H.R. (1999). Developmental microbial ecology of the neonatal gastrointestinal tract. *American Journal of Clinical Nutrition* **69**: 1035S-11045.

Macpherson, A.J. and Uhr, T. (2004). Induction of Protective IgA by Intestinal Dendritic Cells Carrying Commensal Bacteria. *Science* **303**: 1662-1665.

Matsumoto, S., Setoyama, H. and Umesaki, Y. (1992). Differential induction of major histocompatibility complex molecules on mouse intestine by bacterial colonisation. *Gastroenterology* **103**: 1777-1782.

Mazmanian, S.K., Liu, C.H., Tzianabos, A.O. and Kasper, D.L. (2005). An Immunomodulatory molecule of symbiotic bacteria directs maturation of the host immune system. *Cell* **122**: 107-118.

Meyerholz, D.K., Stabel, T.J. and Cheville, N.F. (2002). Segmented filamentous bacteria interact with intraepithelial mononuclear cells. *Infection and Immunity* **70**: 3277-3280.

Moon, H.W. and Bunn, T.O. (1993). Vaccines for preventing enterotoxigenic *Escherichia coli* infections in farm animals. *Vaccine* **11**:213-220.

Noverr, M.C. and Huffnagle, G.B. (2004). Does the microbiota regulate immune responses outside the gut? *Trends in Microbiology* **12**: 562-568.

O'Mahony, C., Scully, P., O'Mahony, D., Murphy, S., O'Brien, F., Lyons, A., Sherlock, G., MacSharry, J., Kiely, B., Shanahan, F. and O'Mahony, L. (2008). Commensal-induced regulatory T cells mediate protection against pathogen-stimulated NF-kappa B activation. *PLoS Pathogen* **4:** e1000112.

Okada, Y., Setoyama, H., Matsumoto, S., Imaoka, A., Nanno, M., Kawaguchi, M. and Umesaki, Y. (1994). Effects of fecal micro-organisms and their chloroform-resistant variants derived from mice, rats, and humans on immunological and physiological characteristics of the intestines of ex-germfree mice. *Infection and Immunity* **62**: 5442-5446.

Osek, J. (1999). Prevalence of virulence factors of *Escherichia coli* strains isolated from diarrheic and healthy piglets after weaning. *Veterinary Microbiology* **68**: 209-217.

Pryde, S.E., Richardson, A.J., Stewart, C.S. and Flint, H.J. (1999). Molecular analysis of the microbial diversity present in the colonic wall, colonic lumen, and cecal lumen of a pig. *Applied and Environmental Microbiology* **65**: 5372-5377.

Rakoff-Nahoum, S., Paglino, J., Eslami-Varzaneh, F., Edberg, S. and Medzhitov, R. (2004). Recognition of commensal microflora by toll-like receptors is required for intestinal homeostasis. *Cell* **118**: 229-241.

Rescigno, M. (2001). Dendritic cells express tight junction proteins and penetrate gut epithelial monolayers to sample bacteria. *Nature Immunology* **2**: 361-367.

Russell, E.G. (1979). Types and distribution of anaerobic bacteria in the large intestine of pigs. *Applied and Environmental Microbiology* **37**: 187-193.

Ryu, J.H., Kim, S.H., Lee, H.Y., Bai, J.Y., Nam, Y.D., Bae, J.W., Lee, D.G., Shin S.C., Ha, E.M. and Lee, W.J. (2008). Innate immune homeostasis by the homeobox gene caudal and commensal-gut mutualism in *Drosphila*. *Science* **319**: 777-782.

Salzman, N. H., De Jong, H., Paterson, Y., Harmsen, H.J.M., Welling, G.W. and Bos, N.A. (2002). Analysis of 16S libraries of mouse gastrointestinal microflora reveals a large new group of mouse intestinal bacteria. *Microbiology* **148**: 3651-3660.

Sharma, R. and Schumacher, R. (1995). Morphometric analysis of intestinal mucins under different dietary conditions and gut flora in rats. *Digestive Diseases and Sciences* **40**: 2532-2539.

Stappenbeck, T.S., Hooper, L.V. and Gordon, J.I. (2002). Developmental regulation of intestinal angiogenesis by indigenous microbes via Paneth cells. *Proceedings of the National Academy of Sciences USA* **99**: 15451-15455.

Strachan, D.P. (1989). Hay fever, hygiene, and household size. *British Medical Journal* **18**: 1259-1260.

Suau, A., Bonnet, R., Sutren, M., Godon, J.J., Gibson, G.R. Collins, M.D. and Doré, J. (1999). Direct analysis of genes encoding 16S rRNA from complex communities reveals many novel molecular species within the human gut. *Applied and Environmental Microbiology* **65**: 4799-4807.

Takahashi, S., Egawa, Y., Simojo, N., Tsukahara, T. and Ushida, K. (2007). Oral administration of Lactobacillus plantarum strain Lq80 to weaning piglets stimulates the growth of indigenous lactobacilli to modify the lactobacillal population. *Journal of General and Applied Microbiology* **53**: 325-332.

Tako, E., Glahn. R.P., Welch, R.M., Lei, X., Yasuda, K. and Miller, D.D. (2008). Dietary inulin affects the expression of intestinal enterocyte iron transporters, receptors and storage protein and alters the microbiota in the pig intestine. *British Journal of Nutrition* **99:** 472-480.

Thompson, C.L., Wang, B. and Holmes, A.J. (2008). The immediate environment during postnatal development has long-term impact on gut community structure in pigs. *ISME Journal* **2:** 739-748.

Umesaki, Y., Setoyama, H., Matsumoto, S., Imaoka, A. and Itoh, K. (1999). Differential roles of segmented filamentous bacteria and clostridia in development of the intestinal immune system. *Infection and Immunity* **67**: 3504-3511.

Umesaki, Y., Setoyama, S., Matsumoto, S. and Okada, Y. (1993). Expansion of αβT-cell receptor-bearing intestinal intraepithelial lymphocytes after microbial colonisation in germ-free mice and its independence from thymus. *Immunology* **79**: 32-37.

Wills-Karp, M., Santeliz, J. and Karp, C.L. (2001). The germless theory of allergic disease: revisiting the hygiene hypothesis. *Nature Reviews Immunology* **1**: 69-75.

Wilson, A.D., Haverson, K., Southgate, K., Bland, P.W., Stokes, C.R. and Bailey, M. (1996). Expression of major histocompatibility complex class II antigens on normal porcine intestinal endothelium. *Immunology* **88**: 98-103.

Wostmann, B.S. (1996). *Germfree and gnotobiotic animal models: background and applications.* Boca Raton, FL: CRC Press.

Zaneveld, J., Turnbaugh, P.J., Lozupone, C., Ley, R.E., Hamady, M., Gordon, J.I. and Knight, R. (2008). Host-bacterial coevolution and the search for new drug targets. *Current Opinion in Chemical Biology* **12**: 109-114.

Zoetendal, E.G., Akkermans, A.D. and De Vos, W.M. (1998). Temperature gradient gel electrophoresis analysis of 16S rRNA from human fecal samples reveals stable and host-specific communities of active bacteria. *Applied and Environmental Microbiology* **64**: 3854-3859.

Zoetendal, E.G., Von Wright, A., Vilpponen-Salmela, T., Ben-Amor, K., Akkermans, A.D.L. and De Vos, W.M. (2002). Mucosa-associated bacteria in the human gastrointestinal tract are uniformly distributed along the colon and differ from the community recovered from feces. *Applied and Environmental Microbiology* **68**: 3401-3407.

Digestive function of amino acids: implications for pig health and performance

Ronald O. Ball
Alberta Pork Research Chair in Swine Nutrition, Department of Agricultural, Food and Nutritional Sciences, University of Alberta, Edmonton, Alberta T6G 2P5, Canada

1. Introduction

In addition to its role in digestion and absorptions of nutrients, the gut is also a major metabolic organ in the body. Although the portal drained viscera (intestines, pancreas, spleen, stomach) only account for 5-7% of body mass, they account for 20-35% of whole body energy expenditure and protein synthesis (McNurlan and Garlick 1980). This review will discuss recent research from our laboratory demonstrating that the gut also extracts a significant portion of the dietary and endogenous amino acids prior to metabolism by the liver and the rest of the body. These gut activities involve the synthesis, interconversion and degradation of amino acids (Bertolo *et al.*, 2002). We have demonstrated, comparing metabolism in pigs fed via the stomach, portal vein or central vein, that the small intestine is even more important than the liver in modifying whole body amino acid utilisation (Bertolo *et al.*, 1999, 2000). Other researchers have also shown that splanchnic tissues may metabolise between 20 and 50% of dietary amino acids (see Bertolo *et al.*, 2004 for a review). Because of this extensive metabolism and the question of whether the gut has an obligatory reliance on dietary, as compared to arterial, amino acids, we have been investigating the small intestine with respect to its impact on whole body amino acid requirements.

This paper will review our recent research on the role of several amino acids (threonine, glutamate, arginine, citrulline, ornithine, methionine/cysteine, branched chain amino acids, lysine) in the small intestine, of piglets before and after weaning, and in growing pigs. The implications of dietary deficiency and of supplementation of these amino acids to swine diets will be discussed.

2. Adequate threonine is critical for gut mucin production

Threonine (THR) is a dietary indispensible amino acid that is usually the second limiting amino acid, after lysine, in wheat and barley grain and is third limiting in typical mixed swine diets. Threonine is found in high concentrations in a number of gastrointestinal secretions that act to protect the mucosa from digestive proteases, prevent the dehydration of the underlying membranes and protect the gut wall from microorganisms and parasitic invasion (Lamont, 1992). Threonine is therefore believed to play an important role in the proper development and functioning of the intestine.

There has recently been increased emphasis on THR metabolism because lysine and methionine, the 1st and 2nd limiting amino acids, have become cost effective for routine supplementation in nearly all swine diets. This has increased interest in the proper inclusion levels of THR in pig diets. This increased interest and demand has also increased competition between companies and subsequently reduced the price of THR to where it is becoming increasingly economical to routinely supplement THR in pig diets.

2.1. The pig intestine uses 60% of dietary threonine

Radioactive tracer techniques and the indicator amino acid oxidation method were used to measure the THR requirement of orally-fed piglets compared to piglets fed intravenously (Bertolo *et al.*, 1998). The purpose of intravenous feeding was to separate the amino acid requirements of the gut (orally fed) from the amino acid requirements of the rest of the body (intravenously fed). The THR requirement of orally-fed piglets was 0.42 g/kg body wt/d and the THR requirement of intravenously fed piglets was 0.19 g/kg/d (Figure 1). Growth rate and final body weight was not different between treatments.

These data showed that the gut of the piglet uses a remarkable 55% of the THR intake. It was confirmed by research by Stoll *et al.* (1998) who used a different method to show that about 60% of dietary THR was taken up on first pass by the pig gut. Threonine must therefore be very important to gut growth, development and function - but what does it do?

Within the gut, THR is found in high concentrations, relative to dietary proteins, in a number of gastrointestinal secretions. In particular, THR (along with serine and proline) is highly concentrated in the mucin proteins of intestinal mucus secretions (Table 1). Because THR is an important constituent

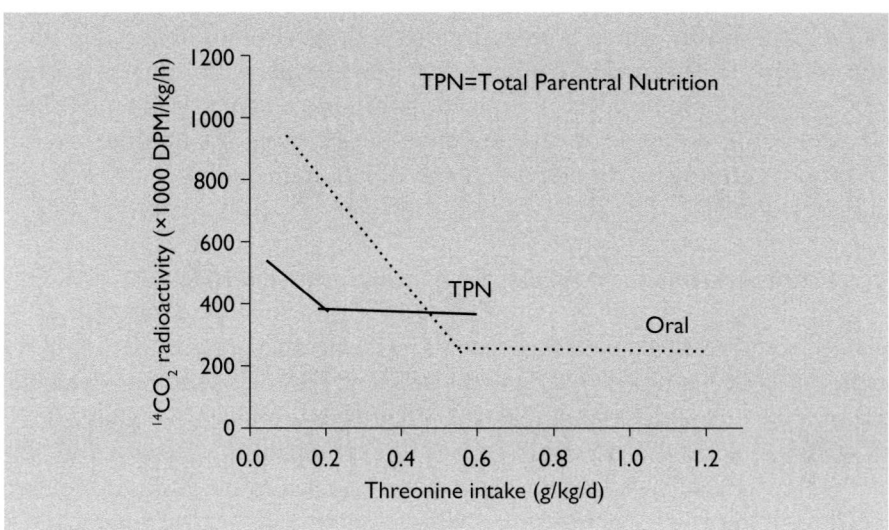

Figure 1. Effect of route of feeding threonine on requirement.

Table 1. Amino acid composition of crude mucin from pigs (Adapted from Lien et al., 1997).

Amino acid	% Crude mucin	SEM (n=3)
Arg	2.4	0.03
His	1.3	0.07
Ile	2.7	0.07
Leu	5.2	0.3
Lys	2.3	0.14
Met	0.6	0.07
Thr	16.4	0.33
Val	6.0	0.06
Ala	9.9	0.63
Asp	7.0	0.34
Glu	8.2	0.29
Gly	8.7	0.65
Pro	12.4	0.69
Ser	12.4	0.74
Tyr	2.1	0.13

of the mucus layer, which lines and protects the gastrointestinal (GIT) tract, the need for THR in mucin synthesis may help to explain the large proportion of dietary THR requirement used by the gut (Bertolo *et al.,* 1998; Stoll, 1998). The correct intake of THR may therefore be critical in the luminal support of proper gut growth, development, and function, including gut health and disease resistance.

2.2. Deficient threonine intake may be associated with diarrhoea

We fed early-weaned piglets an adequate THR diet (0.6 g/kg/d; NRC, 1998), a deficient THR diet (0.1 g/kg/d), and the deficient THR diet with a supplemental intravenous (i.v.) infusion of THR (0.5 g/kg/d THR orally + 0.1 g/kg/d THR i.v.) (Law *et al.,* 2007). The purpose of this experiment was to determine the effect of a THR deficiency on the piglet gut, and to determine if the pig required THR from the diet for gut growth and function or whether the gut could use THR from the blood stream.

As expected, piglets receiving the deficient diet had lower plasma THR and higher plasma urea and lower % nitrogen retention (an indicator of whole body protein deposition) than piglets receiving the adequate intake (Law *et al.,* 2007). There was a small difference in growth rate but the difference was not significant, probably due to higher fat deposition by piglets receiving the deficient diet.

There was a striking difference in incidence of diarrhoea. All pigs in the THR-deficient group exhibited diarrhoea on 35 out of a possible 60 diarrhoea days, with an average severity of 2.82 (out of a maximum of 3.0) over the diarrhoea days (Table 2). Mean daily scores for all pigs in this treatment group over the entire study period was 1.6. Severe diarrhoea was observed by the 4th day of the study and the change to severe diarrhoea was abrupt. In the THR-adequate group, only 1 piglet exhibited diarrhoea over a total of 2 days, with an average severity of 1.5 and an overall mean daily score of 0.05 for all piglets within the THR-adequate group. In the THR-supplemented group, 2 piglets were observed to have slight and moderate diarrhoea, over 2 and 7 days, respectively. The average severity of diarrhoea over the 9 diarrhoea days was 1.3, and the mean daily score for all piglets within the treatment group was 0.2. *Post-mortem* evaluation of the piglets by a veterinary pathologist revealed no evidence of clinical intestinal disease in the THR-deficient pigs. Therefore, the cause of diarrhoea in these pigs was likely due to the effects of THR deficiency on intestinal structure and function.

Table 2. Incidence of diarrhoea in piglets receiving deficient or adequate intake of threonine.

Treatment	# pigs	Mean daily score[1]	# Pigs with diarrhoea	# Diarrhoea days	Average severity[2]
Deficient[3]	6	1.63[a]	6	35	2.82[a]
Adequate	6	0.05[b]	1	2	1.50[b]
Supplemented	6	0.2[b]	2	9	2.21[b]
SD pooled		0.83			0.83

[1] Mean daily score (0 to 3): average of the sum of daily scores per # days of study (Law, 2007).
[2] Average severity score (0 to 3): average daily score on days where diarrhoea was present.
[3] [a,b] denote significance by LSD multiple comparison test ($P<0.05$).

A comprehensive set of intestinal parameters were also measured. The lengths, weights and weight/length ratio of intestine were all lower in the deficient group than in the adequate and supplemented groups. The direction and consistency of response in the measured parameters indicate that intravenously supplied THR did not support gut growth to the same extent as orally supplied THR. We concluded that a diet deficient in THR hindered gut growth even when all other nutrients were available.

Dietary THR intake directly affected mucosal proteins and hence, the weight of mucosal scrapings (Table 3). In each section of the gut examined, the mucosal weight of the deficient group was lower than the supplemented group, which was lower than the adequate group. Therefore, adequate oral intake of THR is required for optimal growth and structure of the mucosa and although intravenous supplementation of THR may be able to make up some of the intestinal requirement, it cannot satisfy it all. This means that adequate first pass uptake of THR is critical for gut growth, structure and function and that the gut has an obligatory requirement for dietary THR that cannot be met by THR provided via the blood supply.

Mucus production was measured by analysis of both mucin content of mucosal scrapings and quantitative histo-chemical techniques by counting goblet cells stained by 3 different methods to determine the type of mucin present. Mucin content of the duodenum was most affected by THR intake (μg Mucin/2cm intestine, adequate 59.6, deficient 11.0, supplemented 46.6, $P<0.05$). The

Table 3. Effect of threonine (THR) deficiency in early weaned pigs on weight(g) of intestinal mucosa.

Intestinal section	THR-adequate	THR-deficient	THR-supplemented	Pooled SD
Duodenum[1]	1.100[a]	0.625[b]	0.914[a,b]	0.035
Prox. jejunum	19.37[a]	11.52[b]	17.61[a,b]	6.29
Ileum	6.41	4.53	5.99	1.69
Colon	3.59[a]	1.78[b]	3.24[a]	1.09

[1] Values in the same row with different superscripts are significantly different: $P<0.05$ (Law, 2007).

number of goblet cells in the colon was found to be greatest in piglets receiving a deficient supply of THR, perhaps suggesting an increased number of stem cells directed to becoming goblet cells. Adequate THR intake resulted in more neutral mucins and more mucin in goblet cells near the villus tip, whereas, when THR was supplemented intravenously, there were more acidic mucins in the crypt, and this effect became more pronounced in the more distal intestine. These data imply that THR from the blood stream is less effective in supporting synthesis of mucins in the villi than luminal THR.

Crypt depth measurements indicated a decrease in intestinal cell proliferation in piglets fed a THR-deficient diet. This was observed in all sections of the gut. Villus height in the mid jejunum and ileum indicated a decrease in intestinal cell differentiation in piglets fed a THR-deficient diet, and these differences were also reflected in villus height/crypt depth ratios. Data from the duodenum exhibited similar trends, although not significantly different. A diet deficient in THR resulted in altered villus architecture and thus probably intestinal function. This clearly indicates an essential role for THR in gut function, absorption, and protection.

2.3. Will anti-nutritive factors in pig feeds stimulate increased intestinal mucin secretion and thus increase threonine requirement of pigs?

Most feeds contain some level of anti-nutritive factors (ANFs), such as insoluble fibres, lignins, tannins, and lectins. Ingestion of these ANFs have been shown in rats to increase mucus secretion; this may increase endogenous threonine losses from the small intestine, and thus THR utilisation. Ileal cannulated pigs

(10 kg, n=42) were randomly allocated to one of 7 casein-based diets with: no supplement (control, CON), 3% lignin (LG), 2% tannin (TN), 2% kidney bean (lectin source, LE), 10% wheat bran (WB), 15% barley (BR), and 10% canola meal (CM), which were fed for 16 d (Myrie *et al.,* 2008). These diets were chosen to represent the intake of each ANF that would be typical of a diet based upon mixed grains. All diets were iso-nitrogenous and iso-energetic with the same total THR content. N balance (5 d) was followed by collection of ileal digesta for analysis of mucins and amino acids. Average daily gain (Table 4) was lower for WB pigs (*P*<0.05) in comparison to CON, TN and CM. WB and BR had lower nitrogen retention (*P*<0.05) compared to CM, TN, LE and CON. Ileum total weight/length (*P*<0.05) and mucosa weight/length (*P*<0.05) were greater in WB pigs. Colon total weight/length (*P*<0.05), colon mucosa weight/length (*P*<0.05), and overall large intestine weight/length (*P*<0.05) were all greater in LG pigs. These data provide clear evidence that the intake of specific ANFs can affect the morphology of the intestine in weaning pigs.

Apparent ileal digestibility of THR and protein were significantly lower than casein control for both wheat bran and barley diets. These data show that high fibre diets increase the loss of THR from the small intestine; this THR is lost to the pig and unavailable for growth. The current correction factor used to estimate endogenous amino acid losses by the pig (NRC, 1998) do not account for the additional THR lost when barley and wheat based diets are used. Additional research is necessary to ensure that pigs fed these types of diets

Table 4. ADG, nitrogen retention and apparent threonine and protein digestibility.

	ADG (kg)	% N retained	Ileal threonine digestibility	Ileal protein digestibility
Casein	0.41[b]	76.4[b]	90.7[b]	92.1[b]
Wheat bran	0.31[a]	71.9[a]	78.9[a]	75.2[a]
Barley	0.34[a]	71.8[a]	82.1[a]	85.3[ab]
Lignin	0.36[ab]	74.6[b]	90.7[b]	91.5[b]
Canola meal	0.41[b]	75.6[b]	85.2[ab]	87.2[b]
Kidney bean	0.37[ab]	77.9[b]	85.1[ab]	86.4[b]
Tannin	0.41[b]	75.9[b]	86.8[ab]	87.1[b]
Mean	0.37	74.8	85.4	86.1

[a,b] Values within columns with different letters are different at *P*<0.05 (Myrie *et al.,* 2008).

are receiving sufficient dietary threonine to account for the additional losses due to anti-nutritive factors in the feeds. Therefore, true ileal digestible THR is probably more dependent upon diet composition than other amino acids and can be expected to vary according to fibre and ANF content of the diet.

To test the hypothesis that dietary THR requirement was affected by feedstuffs, we measured the requirement in pigs fed either casein, which is highly digestible, or barley, which our previous research showed had negative effects on THR digestibility. We found (Myrie *et al.,* unpublished results) that the dietary threonine requirement was higher in the barley fed pigs (Figure 2). These data imply that more THR should be supplemented to diets which increase THR losses, and that the published data on THR digestibility in some feed stuffs may overestimate true THR availability.

2.4. Implications of threonine use by the GIT on pig health and performance

Given the importance of THR in the structure and function of the gastrointestinal tract and the substantial rates of mucin synthesis and secretion (Lien *et al.,* 1997), the dietary requirement for THR is probably increased by conditions that are accompanied by increased mucin secretion and loss from the intestine. These conditions may include: fibre and ANF content of the diet, intestinal

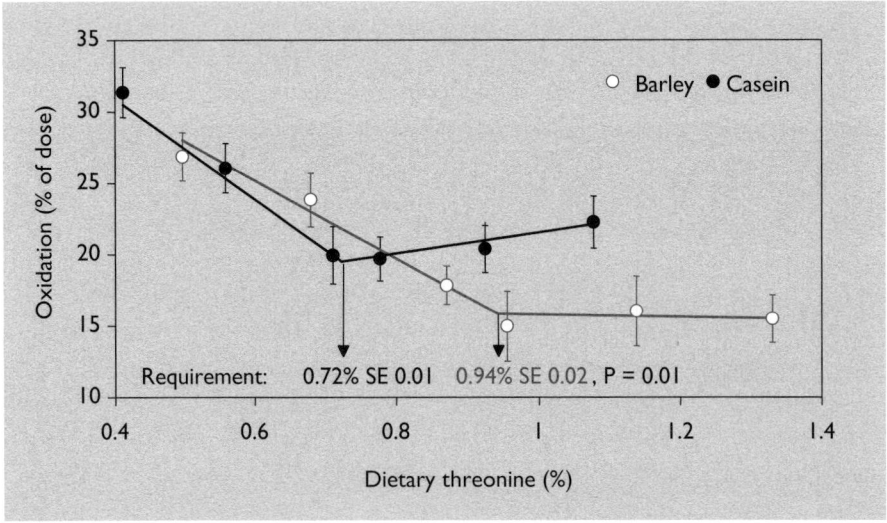

Figure 2. Threonine requirement of weaned pigs fed diets based on casein or barley. (Myrie et al., *unpublished results).*

disease, transition to a new diet, lack of feed intake due to weaning, transport or disease, and during recovery from intestinal disease. Additional dietary THR supplementation during these periods may be beneficial to the maintenance and/or recovery of 'normal' gut function in the pig. Because many feedstuffs increase the excretion of mucins and other endogenous gut proteins due to abrasion, irritation or presence of anti-nutritional factors, inclusion of these ingredients in diets of growing pigs will increase dietary THR requirement due to a higher proportion of dietary THR appearing in endogenous losses. This may leave the pig more susceptible to intestinal upset or disease. THR supplementation in excess of the growth requirement has been reported to increase circulating levels of immunoglobulins (Defa *et al.*, 1999; Wang *et al.*, 2006). Research should be conducted to determine if THR supplementation, in excess of the requirement for growth, has a protective effect against diarrhoea in pigs at risk or if THR supplementation aids or speeds recovery in pigs with intestinal dysfunction. In conclusion, consideration should be given during feed formulation to the potential protective benefits of THR supplementation in excess of the minimum required for growth.

3. Gut utilisation of dietary methionine and cysteine

Stoll *et al.* (1998) measured disappearance of amino acids across the gut of the young weaned pig and found that about 30% of methionine intake was taken up on first pass metabolism by the intestine. To test these results we measured the methionine requirement in piglets fed orally or intravenously, and either with or without cysteine added to the diet. The objective was to determine the requirements of the gut for methionine and whether the gut was essential in the conversion of methionine to cysteine (Table 5) (Shoveller *et al.*, 2003a,b).

Table 5. Methionine (Met) requirements (g/kg per day) of young pigs fed orally or intravenously and either with or without cysteine (Cys) in the diet.

Route of feeding	Met without Cys	Met with Cys (0.55 g/kg per day)	Sparing capacity of Cys
Orally fed	0.42	0.25	40%
Intravenously fed	0.29	0.18	38%

These data show that the intravenous requirement for methionine, without cysteine, is 69% of the oral requirement (0.42 vs. 0.29 g/kg per day). In other words, the gut of the young pig uses 31% of the dietary intake of methionine. The capacity of cysteine to spare the dietary requirement for methionine was similar between orally (40%) and intravenously (38%) fed piglets; this means that the sparing effect of cysteine on methionine requirement is not dependent upon gut function.

The difference in the dietary requirement for methionine between intravenously and orally-fed piglets must be due to uptake of the sulfur amino acids by the small intestine during first pass metabolism. The disappearance of methionine in first pass metabolism by the gut may be due to methionine acting as a methyl donor for synthetic pathways, methionine being converted to cysteine and further metabolites, or by the incorporation of methionine into intestinal proteins (Shoveller *et al.*, 2005). In addition, the incorporation of cysteine into intestinal proteins or the use of cysteine for glutathione synthesis in the gut could also account for the disappearance of sulfur amino acids during first pass metabolism (Ball *et al.*, 2006).

3.1. Implications of intestinal methionine metabolism on gut health

The gut of the young pig uses a significant proportion of the dietary requirement for methionine and thus under conditions of low food intake, diarrhoea or digestive disease, increasing the dietary intake of methionine may be important for maintenance of gut function and metabolism. We are currently engaged in research to identify the functions of sulfur amino acids in the gut of the piglet and the effect of deficiency and supplementation on performance and intestinal function in weaned piglets.

4. The gut may require leucine as an energy source

Using the same approach as previously described for both THR and methionine, we measured the requirements for the branched chain amino acids (BCAA) in orally and intravenously-fed piglets (Elango *et al.*, 2002, 2004). Piglets received diets containing all 3 BCAA in the ratio recommended by NRC (1998) of 1:1.8:1.2 for isoleucine: leucine: valine, respectively. The total BCAA requirement was determined to be 1.47 and 2.61 g/kg/d for intravenous and oral feeding respectively. Thus the intravenous requirement for total BCAA is 56% of the enteral requirement, suggesting that 44% of total BCAA is extracted by first pass splanchnic metabolism. There have been suggestions that the gut uses a large quantity of leucine as an energy source (Elango *et al.*, unpublished results) and this may be the reason for the difference. Previously, the BCAA have been considered to mainly bypass the gut and liver without significant metabolism. Our data and those of others now show that this previous assumption was incorrect. The gut extracts a significant portion of BCAA and leucine catabolism by the gut may actually be required for normal intestinal function.

4.1. Implications of intestinal leucine metabolism on gut health

Our estimate of the BCAA requirement of the young piglet (2.61 g/kg/d) is higher than the current recommended intake by NRC (1998) of 2 g/kg/d. However, there have been very few direct estimates of BCAA requirement in the young pig, as evidenced by the review on amino acid requirements published in the NRC (1998). The current estimate appears to have been derived by extrapolation from data on larger pigs. If the gut of the young pig specifically requires BCAA for energy metabolism or for synthesis of other metabolites, as suggested by our recent data and that of Stoll *et al.* (1998), then the current lack of consideration of the BCAA in formulation of early-weaning piglet diets needs to be re-examined.

5. The gut is the only site of arginine synthesis in the piglet

We have shown that arginine and proline are co-essential amino acids in the diet for the young piglet (Brunton *et al.*, 1999). We found that removal of both arginine and proline from the diet caused greater metabolic disturbances, measured by plasma ammonia levels, compared to when only arginine was removed. When arginine was removed from the diet of young pigs they developed hyperammonemia so severely and so rapidly that they became comatose within hours. Clearly, adequate arginine intake is critical for maintaining normal metabolic functions.

We subsequently demonstrated *in vivo* that the liver and kidney of the young pig lack the complete complement of enzymes to synthesise arginine and proline from other amino acid precursors (e.g. glutamate and glutamine) (Bertolo *et al.*, 2002; 2003). Using multiple radioactive isotope tracer techniques we measured flux, breakdown, synthesis and interconversions of the urea cycle amino acids (Table 6). We showed *in vivo* that the enzyme ornithine amino transferase is only found in the gut of the young pig and is essentially absent from the rest of the body (Bertolo *et al.*, 2003). This means that the piglet requires a fully and

Table 6. Flux and interconversions between arginine, proline and ornithine[2].

Amino acid	Flux (µmol/kg per hour) site of infusion[1]			Percent conversion (%) site of infusion	
	i.g.	i.p.		i.g.	i.p.
Arginine Flux	457±77	292±30[a]	Proline to Ornithine	42±12	1±1[a]
Ornithine Flux	280±46	137±12[a]	Ornithine to Arginine	39±8	11±6[a]
Proline Flux	493±37	426±29[a]	Proline to Arginine	17±7	0±0[a]
			Arginine to Ornithine	53±21	45±11[a]
			Ornithine to Proline	18±6	5±1[a]
			Arginine to Proline	9±4	2±1[a]

[1] Mean ± SD.
[2] Site of infusion: i.g.: intragastric; i.p.: intraportal.
[a] Means differ at $P<0.05$.

normally functioning gut to synthesise arginine and proline, and thus also the urea cycle intermediates that are derived from arginine and proline (Urschel *et al.*, 2007). These data suggest that if the gut of the pig is not functioning normally, or is not receiving enough precursor nutrients, then the pig could rapidly become metabolically-deficient in these amino acids.

Despite providing dietary arginine at about twice the estimated NRC requirement, we showed that the piglets still synthesised an amount of arginine equal to their intake (Wilkinson *et al.*, 2004), which indicates that the actual metabolic requirement for arginine is at least 4 times the estimated NRC (1998) dietary requirement. Therefore, if the pig is not consuming either adequate amounts of arginine and proline, or adequate amounts of the precursor amino acids for synthesis of arginine and proline, then these amino acids could easily become limiting for protein synthesis, intestinal growth and removal of excess ammonia.

Because arginine is used for the synthesis of protein, urea, nitric oxide, and other metabolites, it is possible that the dietary intake to meet the requirement for one for these metabolic activities may not meet the requirement for the others (Ball *et al.*, 2008). For example, does arginine use for ammonia removal via urea synthesis take priority over use for protein synthesis or vice versa?

To answer this question we measured arginine and ammonia concentrations in plasma and muscle protein synthesis rates in piglets receiving intravenous arginine intake varying from 0 to 2 g/kg BW per day (Figure 3 and 4, Brunton *et al.*, unpublished). Intravenous amino acid intake was used so we could by-pass intestinal synthesis and determine the total metabolic demand for arginine by the young pig. Our data showed that: plasma arginine concentration did not become similar to suckled piglets until an arginine intake of about 1.2 to 1.6 g/kg per day; plasma ammonia concentration showed a 2 phase response (rapid decline phase followed by slow decline phase) and was continuing to decline even at an intake of 1.2 g/kg per day; muscle protein synthesis appeared to increase in a linear fashion up to about 1.6 to 2.0 g/kg per day.

5.1. Implications of intestinal arginine metabolism on gut health

These data clearly show that the NRC recommended dietary intake of 0.4 g/kg/d does not meet the total metabolic requirement for arginine by the piglet and therefore the pig must synthesise about 1.2 g/kg BW per day. The implication of this for feeding weaned pigs is that under conditions where the dietary intake of arginine is less than the total metabolic requirement, then a decrease in synthesis rate of arginine due to gut dysfunction could result in metabolic arginine deficiency. A condition when arginine synthesis would be low, is

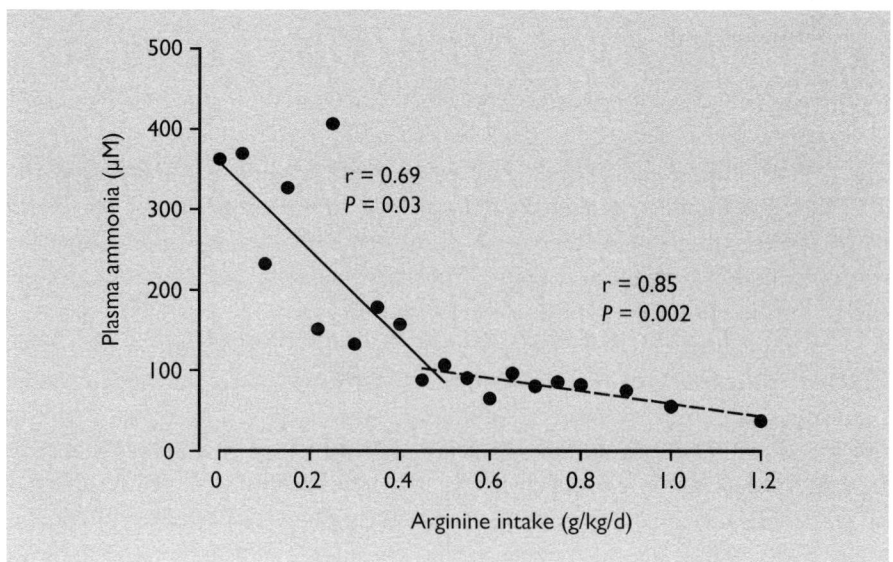

Figure 3. Total metabolic requirement for ARG by plasma ammonia concentration.

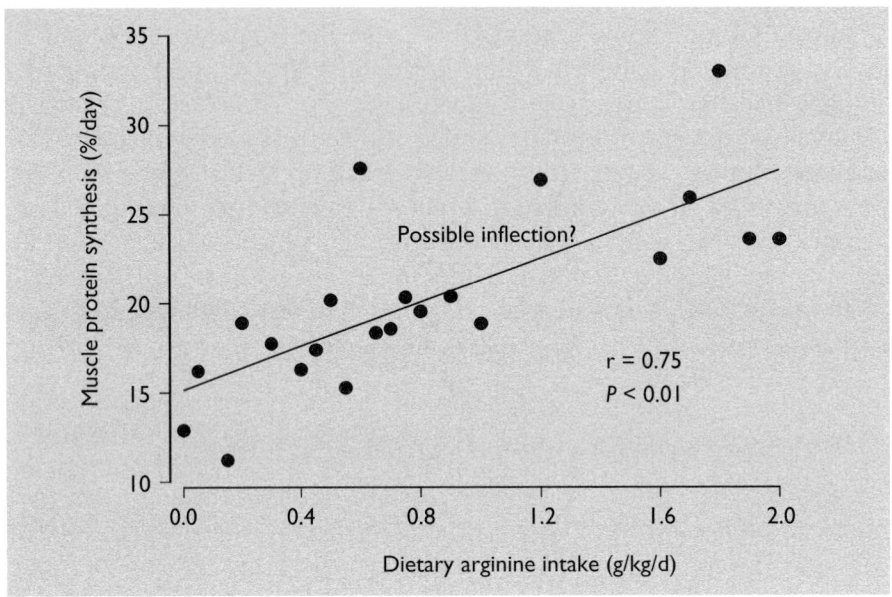

Figure 4. Total metabolic requirement for ARG by muscle protein synthesis.

during low feed intake post weaning, when there may not be enough nutrients and precursors consumed to synthesise adequate amounts of arginine. Research needs to be conducted to determine if pigs with low feed intake and diarrhoea, conditions common at weaning, are able to synthesise sufficient arginine.

6. The pig intestine catabolises lysine

The data of Stoll *et al.* (1998) suggested a significant lysine utilisation by the gut. They subsequently reported a 46% portal balance extraction of lysine and a 22% lysine extraction by the gut. There are very considerable implications for diet formulation and feeding of pigs if the intestine really does catabolise lysine, the 1st limiting dietary amino acid for growth in pigs.

We studied the effects of feeding frequency (Möhn *et al.*, 2003) and lysine intake on lysine oxidation (Moehn *et al.*, 2004). Lysine oxidation in growing pigs at 60, 70, 80 and 90 of individual requirement, determined by the measurement of protein deposition rate of each pig, was measured. The pigs oxidised a considerable portion of their lysine intake, even when it was deficient (Moehn *et al.*, 2004; Figure 5).

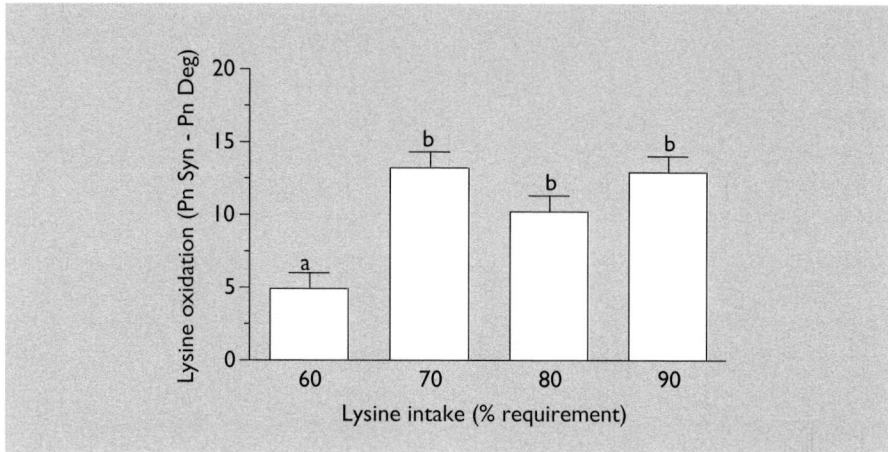

Figure 5. Whole body lysine oxidation at deficient lysine intake (Adapted from Moehn et al., 2004)

We were surprised to find that the literature did not contain any good data on lysine catabolic enzymes in the pig. We found that the published methods, although they worked well in the rat were far from optimal in the pig. We developed new methods, using basic enzymatic and biochemical approaches, for the lysine catabolic enzymes in pig tissues. We have shown (Pink *et al.,* 2003, 2004a,b), in isolated enterocytes, that the small intestine has both the enzymes required and the capacity to catabolise lysine (Figure 6). We confirmed this result by collection of labelled carbon dioxide from enterocytes incubated with ^{14}C-lysine (Figure 7). When activity is expressed per mg/mitochondrial protein, enterocytes appear to have a capacity to catabolise lysine that is about 50% of liver cell capacity.

Although adequate lysine intake is critical to economical pork production there is little information in the literature on the variation in lysine requirement within a population of pigs. Our research clearly indicated that there are differences among pigs in amino acid metabolism. Using the indicator oxidation method we measured individual lysine requirement in a group of pigs (Moehn *et al.,* 2005). Our data confirmed the mean requirement for the population, as published by NRC (1998), and also showed that there is a normal distribution in requirement with a standard deviation of about 10%. This means that feeding to the mean population requirement will underfeed 50% of the pigs! Our new data allows, for the first time, an accurate calculation of the economic impact of over-feeding or under-feeding a population of pigs. This method, which

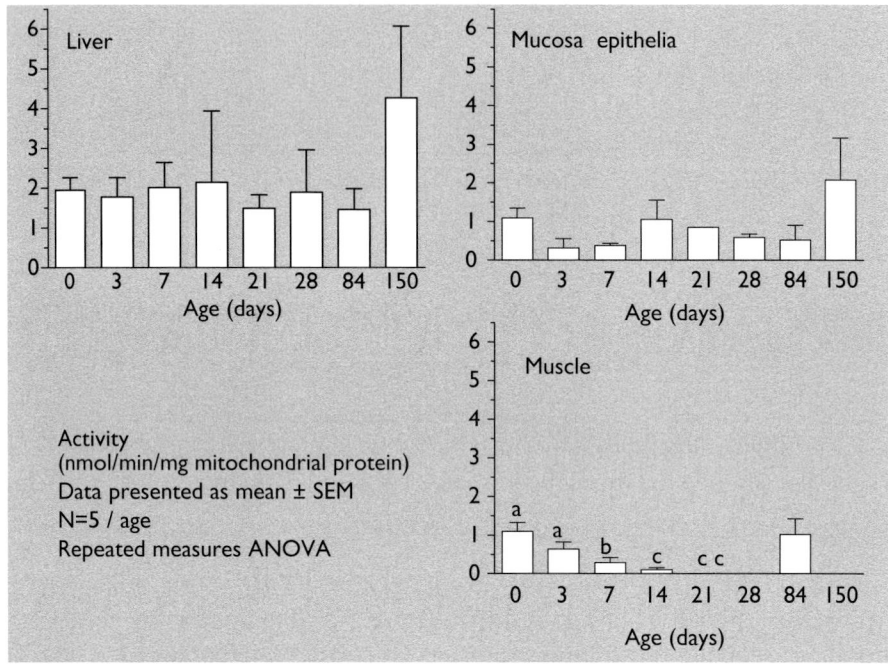

Figure 6. LKR activity in porcine tissues.

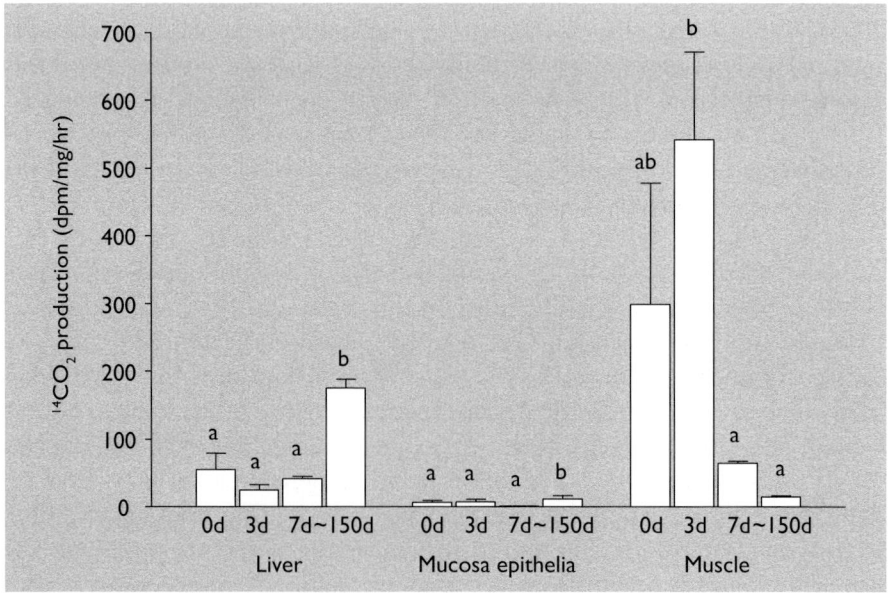

Figure 7. $^{14}CO_2$ production from lysine by porcine mitochondria.

allows the measurement of lysine requirement in individual pigs, also provides the potential to study the reasons for variation in lysine requirement and to genetically select pigs that are more efficient in utilising dietary lysine.

6.1. Implications of obligatory oxidation of lysine

Obligatory oxidation of lysine in both the pig intestine and elsewhere in the body has several important implications. Our data suggest that the intestinal utilisation of lysine is about 10% of whole body requirement and that the intestine uses this amount regardless of the dietary intake. This obligatory whole body oxidation also occurs regardless of whether the intake is deficient or not. One concept this suggests is that there may be an important function to lysine oxidation that we have not yet discovered – why else would the pig continue to catabolise lysine, the most limiting amino acid in the diet, even when intake is deficient? This may also explain why pig growth is so sensitive to lysine intake. Clearly we need to understand a great deal more about the sites, regulation and reasons for lysine catabolism by pigs.

7. Effect of targeted supplementation of early weaning diets with amino acids

As a result of the early experiments on arginine metabolism in the gut of the young pig we carried out an experiment (Ewtushik *et al.*, 2000) in which several amino acids shown to be critical to intestinal development and metabolism were supplemented into a commercial style early weaning diet. The objective of this research was to examine performance and intestinal development in early-weaned piglets receiving diets supplemented with selected amino acids or polyamines. Piglets (3.94±0.43 kg) weaned at 12.5 d were randomly assigned to diets supplemented with either arginine, glutamate, citrulline, ornithine or polyamines, at levels of 0.93, 6.51, 0.94, 0.90 and 0.39%, respectively. Diets were fed for 12 days and growth and intestinal development were measured. There were no significant differences between treatments in performance: average daily feed intake 257 g SD 70; average daily gain 253 g SD 60.

Glutamate supplementation significantly enhanced both total intestinal weight and mucosal growth in several sections of the small intestine, whereas polyamines were detrimental to mucosal growth (Figure 8). Arginine and glutamate supplementation both prevented the weaning-induced reduction in villus height (as a measure of absorptive capacity) in the small intestine (Figure 9), compared to control pigs. Therefore, supplementation of arginine

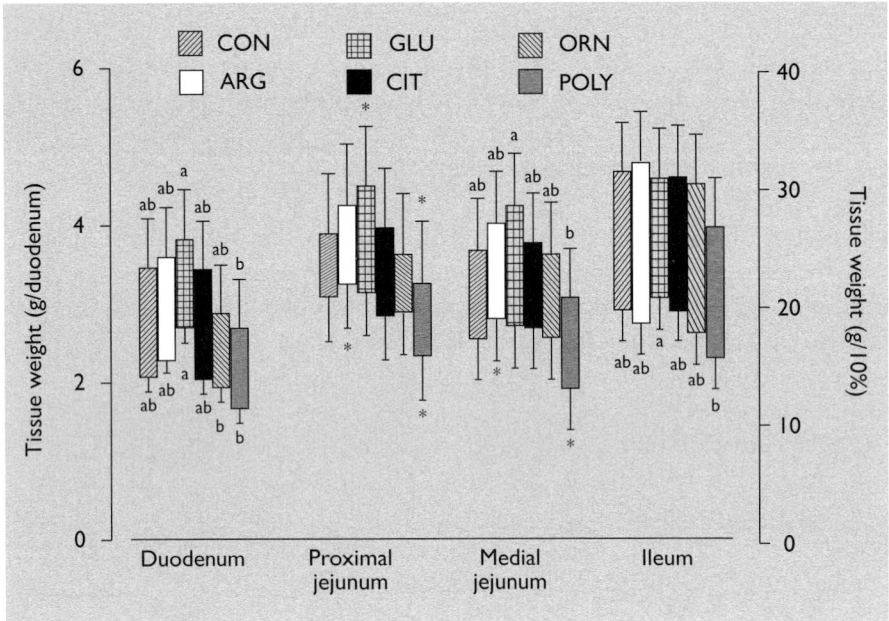

Figure 8. Total and mucosal weights of the small intestine at four sites in piglets fed early-weaning diets.

*Note: control (CON), supplemented with arginine (ARG), glutamate (GLU), citrulline (CIT), ornithine (ORN) or polyamines (POLY). Each bar represents the mean + or - pooled SD for n=7 piglets; bars not sharing the same letter are different (P<0.05); * indicates P<0.10. Open bars - pooled SD represent total mucosal weight per whole duodenum (left axis) or per 10% of total length for respective sections (right axis); whole bars + pooled SD represent total tissue weight per whole duodenum (left axis) or per 10% of total length from respective sections (right axis).*

and its precursor, glutamate, to diets which were presumably more than adequate, based on NRC (1998) recommendations, improved parameters associated with intestinal structure and function. These results indicate that glutamate and arginine supplementation enhance intestinal development of the early-weaned piglet.

Until fairly recently, supplementation of pig diets with glutamate was impractical and uneconomical. However, NuPro™ (Alltech Inc.) is a rich source of highly available amino acids, including glutamate (Mateo and Stein, 2007). The observations described above may provide part of the explanation of why feeding NuPro™ to weaned pigs results in improved performance and animal health (D'Souza and Frio, 2007).

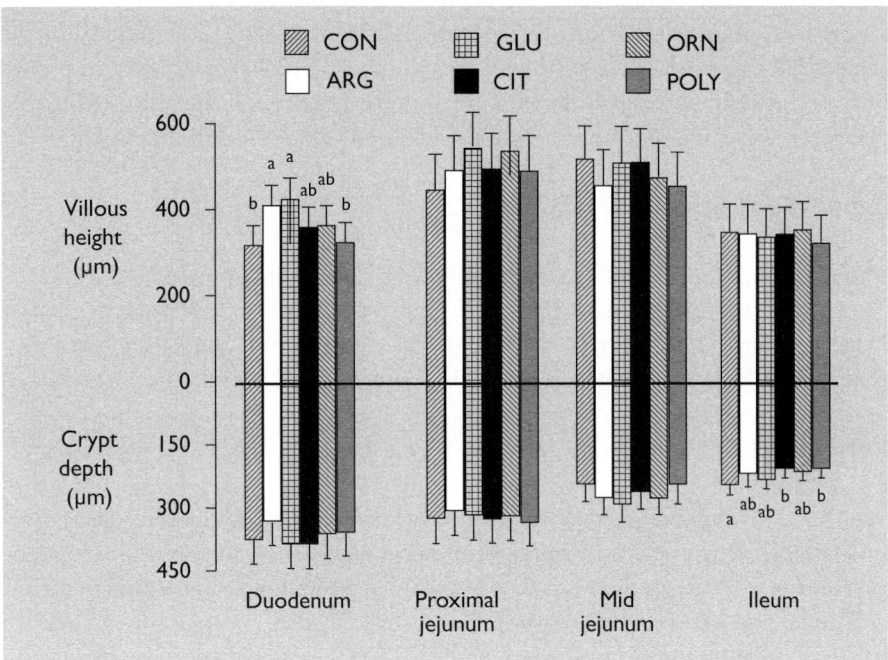

Figure 9. Villus heights (bars above abscissa) and crypt depths (bars below abscissa) of the small intestine at four sites in piglets fed early-weaning diets.
Note: control (CON), supplemented with arginine (ARG), glutamate (GLU), citrulline (CIT), ornithine (ORN) or polyamines (POLY). Each bar represents the mean + pooled SD for n=7 piglets; bars not sharing the same letter are different (P<0.05).

8. Conclusion

In the past, the focus on gut metabolism has been mainly on digestibility of feed ingredients. However, the gut obviously has a major impact on amino acid requirements through specific metabolism and utilisation of individual amino acids. Identification of all the amino acids that are specifically metabolised in the gut is required. In particular, the effect of the gut on metabolism of all the dietary essential amino acids must be further quantified. The effect of feed intake, feed composition, feed additives and digestive diseases on intestinal metabolism and utilisation of amino acids is required to improve our ability to formulate diets for the pig. Increased understanding of how the gut of the pig affects the metabolic availability of amino acids for growth of the gut vs. growth of the rest of the body will lead to improved diets for weaning, specialised diets for recovery from gut stress and intestinal disease, and specialised diets for farms with high prevalence of intestinal disease. In total, these data indicate that supplementation of highly available amino acid to weaning diets, in excess of the requirement for growth, is important for maintenance of gut health and optimum digestive function during this challenging period in the pig's life.

Acknowledgements

The financial support of the following is gratefully acknowledged: Alberta Pork, Alberta Agricultural Research Institute, Alberta Livestock Industry Development Fund, Agriculture and Food Council of Canada, Degussa AG (Germany), Natural Sciences and Engineering Research Council of Canada.

References

Ball, R.O., Courtney-Martin, G. and Pencharz, P.B. (2006). The *in vivo* sparing of methionine by cysteine in sulfur amino acid requirements in animals and adult humans. *Journal of Nutrition* **136**: 168S-1693S. Special Suppl for the 5th International Amino acid Assessment Workshop, Los Angeles, U.S.A.

Ball, R.O., Urschel, K.L. and Pencharz P.B. (2007). Nutritional consequences of interspecies differences in arginine and lysine metabolism. *Journal of Nutrition* **137**: 1626S-1641S. Special Suppl for the 6th International Amino acid Assessment Workshop, Budapest, Hungary. Nov 4-7, 2006.

Bertolo, R.F.P., Chen, C.X.L., Law, G.K.L., Pencharz, P.B. and Ball, R.O. (1998). Threonine requirement of neonatal pigs receiving total parenteral nutrition is considerably lower than that of piglets receiving an identical diet intragastrically. *Journal of Nutrition* **128**: 1752-1759.

Bertolo, R.F.P., Chen, C.Z.L., Pencharz, P.B. and Ball, R.O. (1999). Small intestinal atrophy has a greater impact on nitrogen metabolism than liver bypass in piglets fed identical diets via gastric, central venous or portal venous routes. *Journal of Nutrition* **129**: 1045-1052.

Bertolo, R.F.P., Pencharz, P.B. and Ball, R.O. (2000). Organ and plasma amino acid concentration are profoundly different in piglets fed identical diets via gastric, central venous or portal venous routes. *Journal of Nutrition* **130**: 1261-1266.

Bertolo, R.F.P., Brunton, J.A., Pencharz, P.B. and Ball, R.O. (2002). Arginine ornithine and proline interconversion is dependent on small intestinal metabolism in gastrically fed piglets. *American Journal of Physiology (Endocrinology and Metabolism)* **284**: E915-E922.

Bertolo, R.F.P., Brunton, J.A., Pencharz, P.B. and Ball, R.O. (2003). Arginine, ornithine and proline interconversion is dependent on small intestinal metabolism in neonatal piglets. *American Journal of Physiology (Endocrinology and Metabolism)* **284**: E915-E922.

Bertolo, R.F.P., Ball, R.O., and Pencharz, P.B. (2004). Role of intestinal first-pass metabolism on whole body amino acid requirement. In: (Ed. Burrin, D.G.) *Biology of Growing Animals*. Elsevier.

Brunton, J.A., Bertolo, R.F.P., Pencharz, P.B. and Ball, R.O. (1999). Proline ameliorates arginine deficiency during enteral but not during parenteral feeding in neonatal piglets. *American Journal of Physiology (Endocrinology and Metabolism)* **277**: E223-231.

Defa, L., Changting, X., Shiyan, Q., Jinhui, Z., Johnson, E.W. and Thacker, P.A. (1999). Effects of dietary threonine on performance, plasma parameters and immune function of growing pigs. *Animal feed Science and Technology* **78**: 179-188.

D'Souza, D. and Frio, A. (2007) Bridging the post weaning piglet growth gap: The NuPro®. experience in the Asia Pacific region. Biotechnology in the Feed Industry. 23., 2007, Lexington, Alltech, 2007, pp. 41-48.

Elango, R., Pencharz, P.B. and Ball, R.O. (2002). The branched-chain amino acid requirement of parenterally fed neonatal pigs is less than the enteral requirement. *Journal of Nutrition* **132**: 3123-3129.

Elango, R., Goonewardene, L.A., Pencharz, P.B. and Ball, R.O. (2004). Parenteral and enteral routes of feeding in neonatal piglets require different ratios of branched-chain amino acids. *Journal of Nutrition* **134**: 72-78.

Ewtushik, A.L., Bertolo, R.F.P. and Ball, R.O. (2000). Intestinal development of early weaned pigs receiving diets supplemented with selected amino acids or polyamines. *Canadian Journal of Animal Science* **80**: 653-652.

Lamont, J. (1992). Mucus: The front line of intestinal mucosal defense. *Annals of the New York Academy of Science* **664**: 190-201.

Law, G.K., Bertolo, R.F.P., Adjiri-Awere, A., Pencharz, P.B. and Ball, R.O. (2007). Adequate oral threonine is critical for mucin production and gut function in neonatal pigs. *American Journal of Physiology (Gastrointestinal & Liver Physiology)* **292**: 1293-1301.

Lien, K., Sauer, W. and Fenton, M. (1997). Mucin output in ileal digesta of pigs fed a protein-free diet. *Zeitschrift fur Ernahrungswissenschaft* **36**: 182-190.

Mateo, C.D. and Stein, H.H. (2007). Apparent and standardized ileal digestibility of amino acids in yeast extract and spray dried plasma protein by weanling pigs. *Canadian Journal of Animal Science* **87**: 381-383.

McNurlan, M.A. and Garlick P.J. (1980). Contribution of rat liver and gastrointestinal tract to whole-body protein synthesis in the rat. *Biochemical Journal* **186**: 381-383.

Möhn, S., Fuller, M.F., Ball, R.O. and deLange, C.F.M. (2003). Feeding frequency and type of isotope tracer do not affect direct estimates of lysine oxidation in growing pigs *Journal of Nutrition* **133**: 3504-3508.

Moehn, S., Fuller, M.F., Ball, R.O., Gillis, A.M. and de Lange, C.F.M. (2004). Growth potential, but not body weight or moderate limitation of lysine intake, affects inevitable lysine catabolism in growing pigs. *Journal of Nutrition* **134**: 2287-2294.

Moehn, S., Bertolo, R.F.P., Pencharz, P.B. and Ball, R.O. (2005). Estimate of the variability of lysine requirement of growing pigs using the indicator amino acid oxidation technique. *Journal of Animal Science* **83**: 2535-2542.

Moehn, S., Shoveller, A.K., Rademacher M. and Ball, R.O. (2008). An estimate of the methionine requirement and its population variability in growing pigs using the indicator amino acid oxidation technique. *Journal of Animal Science* (in press).

Myrie, S.B., Bertolo, R.F.P., Sauer, W.C. and Ball, R.O. (2008). Effect of common anti-nutritive factors and fibrous feedstuffs in pig diets on amino acid digestibility with emphasis on threonine. *Journal of Animal Science* **86**: 609-619.

National Research Council (1998). Nutrient Requirements of Swine. 10th revised edition, National Academy of Science, Washington, DC.

Pink, D.B.S., Elango, R., Dixon, W.D. and Ball, R.O. (2003). Intestinal catabolism of lysine in swine. In: *Procedings of the* 9th *International Symposium on Digestive Physiology in Pig*, Banff, Alberta May 14-17, 2003, p. 131.

Pink, D.B.S., Dixon, W.D. and Ball, R.O. (2004a). Lysine catabolism in swine: an enzymatic approach. Abstract #214.7 Experimental Biology Annual Meeting, Washington, DC April 20-24, 2004.

Pink, D.B.S., Elango, R., Dixon, W.D. and Ball, R.O. (2004b). Regulation of lysine degradation during the postnatal stages of growth and development in the pig. *Advances in Pork Production* **15**: A20.

Shoveller, A.K., Brunton, J.A., Pencharz, P.B. and Ball, R.O. (2003a). The methionine requirement is lower in neonatal piglet fed parenterally than in those fed enterally. *Journal of Nutrition* **133**: 1390-1397.

Shoveller, A.K., Brunton, J.A., House, J.D., Pencharz, P.B. and Ball, R.O. (2003b). Dietary cysteine reduces the methionine requirement by an equal proportion in both parenterally and enterally fed piglets. *Journal of Nutrition* **133**:4215-4224

Shoveller, A.K., Ball, R.O., Stoll, B., and Burrin D. (2005). Nutritional and functional significance of intestinal sulfur amino acid metabolism. *Journal of Nutrition* **135**:1609-1612.

Stoll, B., Henry, J., Reeds, P., Yu, H., Jahoor, F. and Burrin, D. (1998). Catabolism dominates the first-pass intestinal metabolism of dietary essential amino acids in milk protein-fed piglets. *Journal of Nutrition* **128**: 606-614.

Urschel, K., Rafii, M., Pencharz, P.B. and Ball, R.O. (2007). A multi-tracer stable isotope quantification of the effects of arginine intake on whole body arginine metabolism in neonatal piglets. *American Journal of Physiology (Endocrinology and Metabolism)* **293**: E811-E818.

Wang, X., Qiao, S.Y., Liu, M. and May, Y.X. (2006). Effects of graded levels of true ileal digestible threonine on performance and serum parameters and immune function of 10-25 kg pigs. *Animal Feed Science and Technology* **129**: 264-278.

Wilkinson, D.L., Bertolo, R.F.P., Brunton, J.A., Shoveller, A.K., Pencharz, P.B. and Ball, R.O. (2004). Arginine synthesis is regulated by dietary arginine in the enterally fed neonatal piglet. *American Journal of Physiology (Endocrinology and Metabolism)* **287**: E454-E462.

Nutritional maintenance of gut health: an impending reality?

Bernard Sève, Alice Hamard and Nathalie Le Floc'h
INRA-Agrocampus Rennes, Unité Mixte de Recherches Systèmes d'Elevage, Nutrition Animale et Humaine, 35590 Rennes, France

1. Introduction

Maintenance of gut health in the pig is probably the last emerging problem when the other pathologies are under control. Gut health degradation is most often observed during the post-weaning period and may in turn affect further general health through enhancement of bacterial translocation and disturbances in the maturation of the immune system including immuno-tolerance. Antibio-supplementation has been first proposed for controlling weaning scouring and preventing diarrhoea. Since the ban on this practice in Europe, several classes of alternatives have been developed: organic acids, probiotics, immunoglobulins (e.g. from Spray-dried plasma or bovine colostrum), antimicrobial or immunomodulating substances (e.g. from plant extracts and natural substances PENS), various bioactive or regulatory molecules (e.g. from bovine colostrum or PENS) and prebiotics or fermentable fibre from the various feed ingredients. As nutritionists we have been concerned by alternatives aimed at optimising feed intake according to feed ingredients and balances between nutrients. Evidently, the best weaning diet would be one using highly digestible feedstuffs simulating the mother's milk. However, this is not only expensive but will retard the necessary adaptation to a conventional feed. Recommendations of feed rationing or lower dietary protein content, have been shown to preserve health in reducing growth performance not only in the short but also in the long-term. Finally, with conventional feeds we proposed restricted feeding of a diet with a high protein:energy ratio. In this paper, I will first examine the improvement of this concept through the proposal of using fermentable dietary fibre in order to optimise the protein/carbohydrate balance of fermentations in the gut. Then, I will examine possible specific requirements for amino acids involved not only in the preservation of the expression of muscle growth potential in the long-term, but also in the preservation of gut functionality and health in the short-term. Then I will conclude on the need for more knowledge in the dietary control of pathogens and the other specific requirements for nutrients involved in gut health.

2. Alternatives to antibiotics: the use of high fibre feed ingredients or prebiotics

In France, the use of barley instead of wheat or even corn in diets for early-weaned pigs has been recommended for many years (Aumaitre, 1969, 1975; Quéméré *et al.*, 1977). The main observations were reduced weaning diarrhoea and, despite lower digestible energy and protein supplies due to a comparable dry matter intake of lower energy value, similar or higher growth performance immediately post-weaning (Table 1). This might have been related to a beneficial action of barley fibre on the digestive physiology and (or) the gastro-intestinal microbiota of the piglet.

However, barley is a composite fibre source containing both soluble and insoluble non-starch polysaccharides. Literature data indicate that soluble fibre from the endosperm made of soluble mixed-linked beta-glucans may increase viscosity in the stomach and in the small intestine lumen. The rate of gastric emptying was shown to be decreased with dietary fibre, particularly those increasing viscosity such as sugar beet pulp in sows (Miquel *et al.*, 2001). In weaned piglets, it was shown that an abrupt switch from a milk-based to a cereal-based diet transiently accelerated gastric emptying with wheat but normalised and then decreased gastric emptying rate with barley (Boudry *et al.*, 2004, Figure 1). There are reasons to think that the rate of passage is

Table 1. Impact of barley vs. other cereal grains on growth performance and gut health in weanling piglets.

Cereal	Wheat	Maize	Barley	References
Feed intake, g/d	585[a]	597[a]	629[b]	Aumaitre *et al.*, 1969
Daily gain, g/d	360[a]	354[a]	386[b]	
FCR	1.61	1.64	1.59	
Diarrhoea, days	3.4[a]	3.5[a]	1.6[b]	
Feed intake, g/d	578	572	600	Quéméré *et al.*, 1977
Daily gain, g/d	347	344	333	
FCR	1.68	1.67	1.82	
Diarrhoea, n	32[a]	24[a]	12[b]	

Note: means in a row with superscripts without a common letter differ, $P<0.05$.

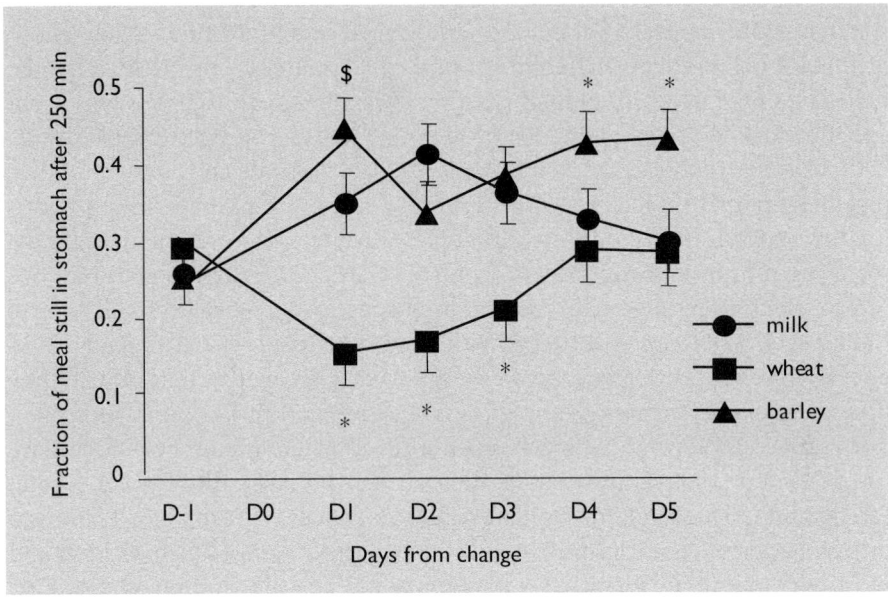

Figure 1. Effect of a sudden change from a milk to a cereal grain based diet on gastric emptying. The stars indicate significant differences (P<0.05) between wheat and barley data (Boudry et al., *2004).*

slowed in the small intestine and that this will favour the development of the microbial flora there. The Australian group of Hampson (McDonald *et al.,* 1999, 2001) found that this increase in viscosity caused either by guar gum or non-fermentable carboxymethylcellulose (CMC) favoured the implantation of haemolytic enterotoxaemic *Escherichia coli* (ETEC) and deteriorated gut health status. This was shown also with pearl barley in which the soluble NSP fraction is maximised in regard of the insoluble one localised in hulls (Hopwood *et al.,* 2004). According to the authors, although beta-glucans are fermentable fibre, the undigested starch due to the immaturity of the digestive enzymes was preferentially used by the microbiota. What is surprising is why the usually dominant flora of lactobacilli in the small intestine (Konstantinov *et al.,* 2004; Hill *et al.,* 2005) could not prevent proliferation of pathogenic enterobacteriacae like ETEC, despite a trend to a decrease in luminal pH.

2.1. Interaction of fibre with protein supply for gut health

A possible explanation may be that during the first two weeks following weaning, due to the immaturity of the enzymatic systems, there is not only undigested starch available at ileal level but also undigested protein of either endogenous or dietary origin favouring coliform bacterial proliferation. We investigated the effect of the incorporation of CMC in a weaning diet for piglets where not only the carbohydrate fraction but protein also was highly digestible (Piel *et al.*, 2005). Despite highly significant increases in ileal viscosity and faecal humidity, piglets remained perfectly healthy. Faecal humidity resulted from the water-holding capacity of CMC and, instead of favouring hypersecretion, reduced the secretory response to cholinergic agonist (Lallès *et al.*, 2006). The deleterious impact of high/low-digestible protein supply on piglet gut health is well known. Bikker *et al.* (2006) investigated the interaction between dietary fibre and level of conventional protein sources in diets for early-weaned piglets. During the first 2 weeks after weaning, all piglets remained healthy, but the high protein diet reduced performance, which was improved with a mixture of soluble and insoluble fibre sources and low-digestible starch. Although the fibre per protein interaction was not significant, fibre was shown to reverse the ratio of coliform to lactobacilli bacteria, stimulating lactic acid production at jejunal level and reducing ammonia contents at both jejunal and colonic level. In addition, an elevation of colonic butyrate was found in the colon, which could be beneficial to the development and functionality of the large intestine.

On the other hand, fibre itself was shown to decrease the apparent or standardised digestibility of protein in pigs. Using protein-free diets this appeared to be caused by an increase in non-recycled endogenous protein from mucus, enzymes or sloughed intestinal cells (De Lange *et al.*, 1989) although this increase could plateau above 3.4% of crude fibre from mixed fibre sources (Mariscal-Landin *et al.*, 1995) at a level close to that of basal endogenous losses (Hess *et al.*, 2000). However, measurements of ileal endogenous protein losses (IEPL) through isotope-labelled nitrogen or amino acid methods revealed the existence of additional specific IEPL directly related to the protein content of feedstuffs. There was some evidence of a positive interaction between protein and fibre leading within each type of protein source to positive regressions of endogenous losses on fibre content (Sève and Lahaye, 2003; Figure 2). Viscosity or water retention capacity (WRC) was shown to increase ileal endogenous protein losses (Larsen *et al.*, 1993). This appeared to be particularly true with partly purified pea cotyledon fibre (Leterme *et al.*, 1998) made of insoluble fibre with high WRC, although comparing entire peas from cultivars differing in pea cotyledon fibre contents we did not find any relationship with IEPL (Hess *et al.*, 1998). Similarly, we compared the endogenous losses induced by entire seeds of wheat and barley of different beta-glucan contents and viscosity (Skiba *et al.*, 2007). The ileal endogenous protein losses were significantly higher with barley

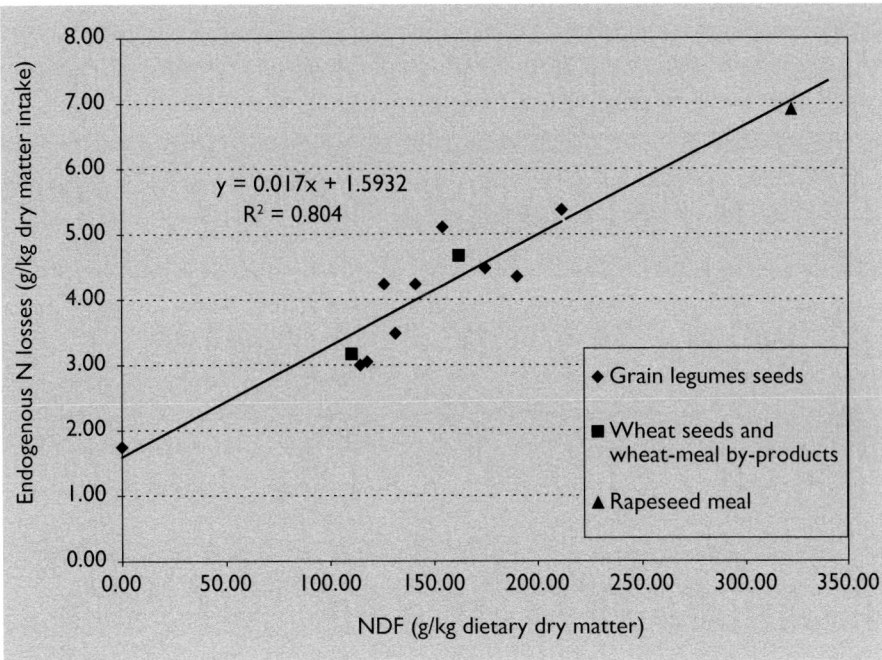

Figure 2. Variation of N endogenous losses according to feedstuff NDF content. Measurement by isotopic dilution method with 15N-labeled diets (Sève and Lahaye, 2003).

than with wheat but better explained by total fibre content of the seeds than by their content in soluble or insoluble beta-glucans, or their own viscosity. In fact, contrary to findings with pearl barley (Hopwood *et al.*, 2004), the viscosity of the ileal contents remained low and unrelated to the ileal flows of dietary or endogenous protein. This would support the conclusive statement of Bach Knudsen (2008) that 'a mix of soluble and insoluble fibre and carbohydrate components could prove to be beneficial'.

Nevertheless, the claimed 'specific stimulation of lactobacilli bacteria over enteropathogenic bacteria' of the latter author might strictly depend on the balance between residual fibre or resistant starch and undigested endogenous or dietary protein to be transferred from the ileum to the caecum. We recently tested this hypothesis at constant dietary protein level, in the presence of pea as a low-digestible conventional protein and natural fibre sources of different solubility (Table 2). Measured one week after weaning, the apparent digestibility of protein was low but much less severely depressed than expected from calculations

Table 2. Impact of the enrichment of a pea-based diet in insoluble or soluble fibre, on ileal amino acid endogenous losses and true digestibility, and on nitrogen balance in piglets at second week after weaning (Sève et al., unpublished data).

Diet	Control low fibre [1]	Pea, low fibre [1]	Pea, high fibre + insoluble [1]	Pea-high fibre + soluble [1]
Total fibre, %	10.9	13.2	19.2	19.5
% soluble	26.7	23.5	17.4	34.4
Feed intake, g/d	300[a]	304[a]	287[a,b]	264[b]
Weight gain, g/d	290	301	301	311
Feed/gain	0.97[a]	1.00[a]	1.06[a,b]	1.18[b]
10 amino acids				
Apparent digestibility, %	66.3[a]	53.3[b]	63.5[a]	65.1[a]
Total endogenous loss,%	11.2[a]	11.6[a]	12.3[a]	13.8[a]
True digestibility, %	77.6[a]	64.9[b]	75.7[a]	78.9[a]
4 essential amino acids				
Apparent digestibility, %	72.6[a]	54.9[b]	64.5[a]	66.6[a]
Total endogenous loss, %	8.0[a]	8.0[a]	8.4[a]	10.2[a]
True digestibility, %	80.6[a]	62.9[b]	73.0[a]	76.9[a]
Ileal mucin flow, g/kg dry matter intake	24.7[a]	32.1[a,b]	27.7[a]	34.7[b]
Caecal pH	5.74[a]	5.64[a]	5.50[a,b]	5.26[b]
Caecal ammonia, mmol /g	218	212	128	79
Caecal lactic acid, mmol /g	12.6[a]	1.6[b]	9.4[a,b]	23.0[a]
Clostridium Perfringens in proximal colon (% of control)	100	85	25	0

[1] Means in a row with superscripts without a common letter differ, P<0.05.

assuming additivity of indigestible protein from each separate feedstuff, when pea was associated with insoluble or soluble fibre. The numerical increase in endogenous protein losses with soluble fibre was not significant and true amino acid digestibilities of the high fibre diets matched values for the control low-fibre low-pea diet. Compared to a low-digestible protein without fibre addition, increasing dietary fibre with soluble fibre in the form of sugar beet pulp (SBP) significantly decreased pH and ammonia and increased lactic acid content in the caecum. Increasing dietary fibre with insoluble fibre mainly in the form of wheat bran gave intermediate values. This suggested, as anticipated, that lactobacilli bacteria were stimulated (Konstantinov *et al.*, 2004) and evidence for the inhibition of pathogenic bacteria development was provided by the fact that *Clostridium perfringens* could not be detected anymore in the proximal colon of piglets fed the SBP diet. Remarkably with this diet, despite a trend to lower feed intake probably due to higher WRC responsible for the bulky consistence of the meal, feed efficiency and nitrogen utilisation were improved compared to the two other high-pea diets matching the performance achieved with the control diet.

It is now well known that fasting and underfeeding following weaning result in mucosal villi atrophy (Carey *et al.*, 1994) with reduction of functional, i.e. digestive and absorptive, capacity (Nunez *et al.*, 1996; Boudry *et al.*, 2002). Fasting and underfeeding also negatively affect the barrier function of the intestine increasing the risk of toxins absorption or bacterial translocation (Carey *et al.*, 1994; Boudry *et al.*, 2004). Other effects of underfeeding have been reported on goblet cells counts and mucin contents (Nunez *et al.*, 1994), mucosal glutathione concentrations and Peyer's patch lymphocytes counts (Lopez-Pedrosa *et al.*, 2007). The question of the most limiting nutrients for an adequate development of the mucosa as well as its barrier and immune function is a highly active area of research.

3. Glutamine, arginine and proline as conditionally essential amino acids involved in gut health

A very severe protein restriction (90%) affected the growth of digestive tissues (Dudley *et al.*, 1997) but, with less marked reduction of a balanced protein supply (50-66 %), this development was at least partially preserved at the expense of carcass tissues, particularly muscle, through a post-weaning increase in protein synthesis rate regardless of dietary protein level (Sève *et al.*, 1986; Ebner *et al.*, 1994). Nevertheless, the acceleration of protein synthesis may be insufficient to fully compensate for the simultaneous increase in protein degradation (Adegoke *et al.*, 2003) eventually caused by the lower availability of luminal amino acids (Figure 3), inducing a decrease in protein accretion (Dudley *et al.*, 2001). Furthermore, it was shown that normal growth and development of the intestinal mucosa cannot be obtained with parenteral nutrition (Bertolo *et al.*, 1999) which induces a decrease in protein synthesis rate (Dudley *et al.*, 1998; Stoll *et al.*, 2000). This suggests a close connection between tissue growth and functionality and draws the attention to those dietary amino acids playing critical physiological roles other than their incorporation into protein.

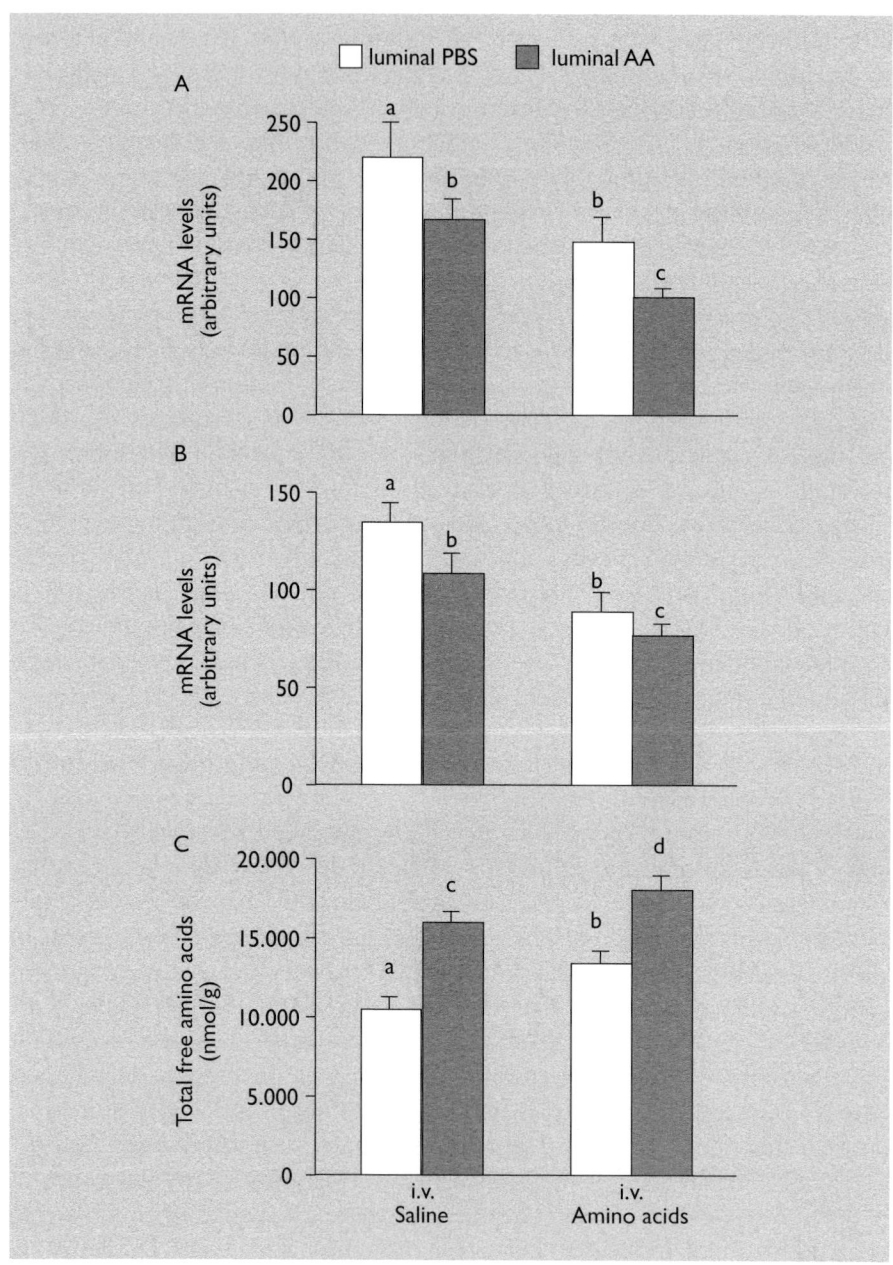

Figure 3. Additive effects of luminal and systemic amino acids on the mucosal expression of the genes for (A) ubiquitin and (B) ubiquitin-conjugating enzyme involved in the control of protein breakdown, and (C) mucosa free AA concentration (Adegoke et al., 2003). Data are means ± SE. Means without a common letter differ (P<0.05; n= 6).

Glutamine is a conditionally-essential amino acid and the preferred energy substrate for rapidly proliferating cells such as those in lymphoid tissues and in the small intestine mucosa. It was shown, mainly with *in vitro* experiments (Le Bacquer *et al.,* 2001) or with *in vivo* administration to humans of high doses (Coeffier *et al.,* 2003), an increase in enterocyte protein synthesis rate. Le Bacquer *et al.* (2003) found that glutamine administered either apically or basolaterally increased protein synthesis on Caco-2 cell monolayer submitted to luminal fasting and this required deamidation to glutamate presumably providing the required energy supply. Enteral glutamate directly (Reeds *et al.,* 1997), and presumably glutamine after deamidation, are also precursors for glutathione synthesis and play an important role in the enteric antioxidant system. However, the effect on protein synthesis *in vivo* remains controversial (Marchini *et al.,* 1999) and cysteine may often be first-limiting for glutathione protein synthesis. Nevertheless, there are indications of a reduction of protein catabolism through the ubiquitin-dependent pathway (Adegoke *et al.,* 2003), increased cell proliferation (De Marco *et al.,* 1999) and reduced apoptosis (Papaconstantinou *et al.,* 1998) with glutamine. Accordingly, the trophic effect of glutamine on the intestinal mucosa in terms of prevention of villus atrophy in piglets post-weaning has been clearly established (Wu *et al.,* 1996; Ewtushik *et al.,* 2000; Domeneghini *et al.,* 2006; Table 3). Most interestingly, in terms of gut health, glutamine was shown to prevent acute endotoxin-induced increase in permeability (Dugan and McBurney, 1995) in piglets. Le Bacquer *et al.,* (2003) got the same result with luminal fasting induced Caco-2 cell monolayer permeability, presumably through regulation of tight junction proteins (Li *et al.,* 2004). Last but not least, glutamine can modulate the inflammatory response decreasing pro-inflammatory and increasing anti-inflammatory cytokine gut production (Coeffier *et al.,* 2001).

Table 3. Effect of a supplement of 0.5% of L-glutamine to a standard weaning diet on ileal mucosa structure. There was no effect on weight gain (Domeghini et al., 2006).

Diet	Control	+ Glutamine	*P<*
Villus height, µm	168	208	0.004
Crypt depth, µm	110	166	0.005
Villus:Crypt ratio	1.53	1.25	0.022
Apoptosis:Mitosis index	0.069	0.062	0.002

Arginine also is a conditionally essential amino acid in pigs that play important physiological roles related to gut and whole-body health. Endogenous synthesis is insufficient to cover the requirement for maximal growth particularly in suckled pigs so that arginine appears to be the first limiting amino acid in sow's milk as well as in milk replacers for 7-21 d-old piglets (Wu *et al.*, 2004). These authors reported that this limitation could be overcome not only through supplementation but also through enhancement of citrulline production from proline or glutamine in the intestinal mucosa mitochondria, and transfer to the cytosol and extracellular space for further conversion to arginine in situ and in the kidney, respectively. In the same way, ornithine produced in enterocyte mitochondria can be transferred to the cytosol and used as a precursor for the synthesis of polyamines required for the stimulation of epithelial cell proliferation and trophicity (Wu *et al.*, 2000). These authors have emphasised the role of physiological levels of glucocorticoids in the enhancement of glutamine-proline-arginine interconversion, not only for whole-body arginine homeostasis (Wu *et al.*, 1997) but also in the maturation of gut metabolism and function after weaning. The lack of maturity of this system is responsible for the higher requirement not only for the conditionally-essential amino acids glutamine and arginine, but also for proline in neonatal piglets (Ball *et al.*, 1986; Samuels *et al.*, 1989).

Arginine is the precursor for nitric oxide (NO) which plays utmost important roles in gut physiology. As a neurotransmitter of inhibitory nonadrenergic noncholinergic neurones, NO was thought to mediate an effect of intragastric arginine administration on normalisation of endotoxin-induced hypermotility in pigs, although an inhibitory effect of insulin could not be excluded (Bruins *et al.*, 2004). As a signalling molecule, NO is involved in the stimulation of enterocyte migration towards a mechanical damage area (Rhoads *et al.*, 2004) and in the inhibition of pro-inflammatory cytokines production in a model of temporary intestinal ischemia, reducing damage to the mucosa (Spanos *et al.*, 2007). It is also involved in angiogenesis through secretion of vascular endothelial growth factor (VEGF). Moderate dietary supplementation of arginine in piglets (0.7%) was recently shown to enhance small intestine microvascularisation simultaneously increasing villus height (Zhan *et al.*, 2008) while higher supplements reversed the NO/endothelin-1 balance with opposite effect (Table 4; Figure 4). In the latter cases, an action of arginine through polyamine production was not excluded. Using transgenic mice overexpressing arginase, arginine was shown to be directly involved, i.e. independent of NO production, in the development of ileal Peyer patches and in the maturation of B lymphocytes.

Table 4. Effects of dietary L-arginine supplementation on growth performance and diarrhoea incidence 10 days following weaning at 2 weeks of age (Zhan et al.. 2008).

	L-arginine supplement [1]		
	0 (control diet)	0.7%	1.2%
Feed intake, g dry matter /kg body weight	30±1	30±3	28±2
Weight gain, g/d	89±12	92±8	69±8
Gain per feed	0.55±0.07	0.59±0.05	0.48±0.06
Diarrhoea incidence, % of 10d-period	8.8[a]	3.8[b]	15[a]

[1] Values are means ± SEM, n=8. Means in a row with superscripts without a common letter differ, $P<0.05$.

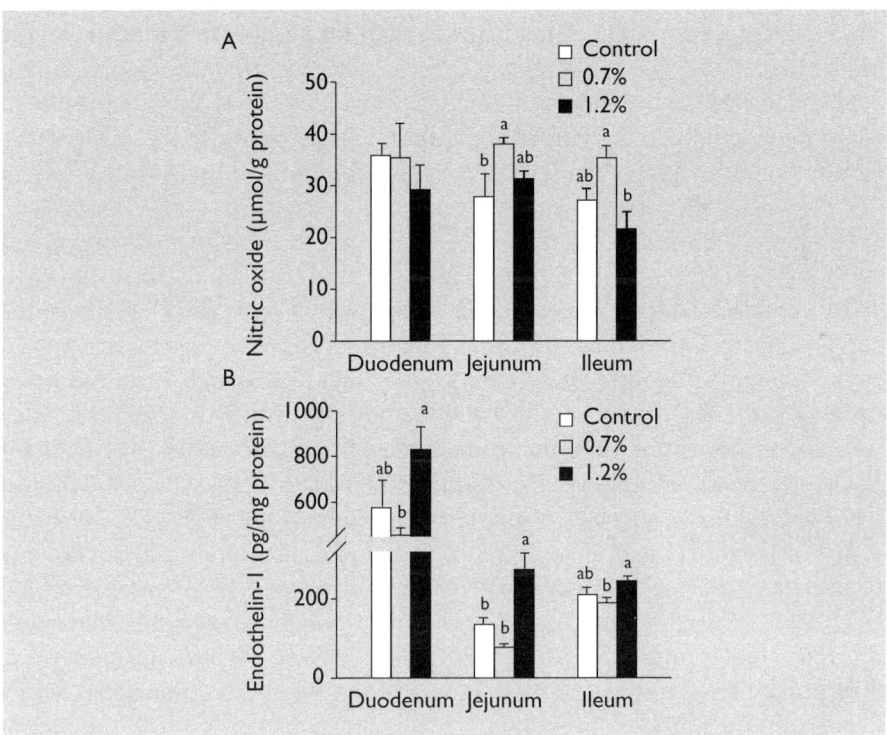

Figure 4. Effects of dietary L-arginine supplementation on nitric oxide and endothelin-1 concentrations in the small intestinal tissue of piglets (Zhan et al., 2008). Bars represent the mean ± SEM, n=6. Within an intestinal region, means without a common letter differ, P<0.05.

Finally, arginine appears not only to control the consequences of digestive infections or insults but also to enhance the development of vascular and immune functionality of the intestinal mucosa in young animals. Altogether, these features probably explain why arginine could reduce the symptoms of necrotising enterocolitis in premature infants (Amin *et al.*, 2002). However, in animals as in human, the use of arginine supplements should be used with caution. In addition to disturbance of physiological balance, chronic dietary supplements over recommendation for piglets may create nutritional imbalance affecting the absorption and tissue utilisation of lysine as reminded by several authors (e.g. Zhan *et al.*, 2008).

4. Threonine

Threonine is a strictly indispensable amino acid, which implies that functions in which it is involved depend on dietary supply. According to different authors, 38-89% of the digested threonine does not appear in the portal vein (Hamard, 2008; Schaart *et al.*, 2005). This is lower than for glutamate but substantially higher than the 17-51% range reported for the other strictly indispensable amino acid lysine. This is the most convincing evidence for a specific requirement of portal-drained viscera (PDV) for threonine and probably the explanation for the higher maintenance requirement for threonine than for lysine which is more involved in the growth of carcass tissues. Regarding PDV extraction, the first hypothesis is an irreversible loss through oxidative pathways, but the pancreas is the only PDV with TDG activity and total oxidation through TDG is quite low (Le Floc'h and Sève, 2005). It was shown that threonine supplements could increase immunoglobulin synthesis in pigs treated with a swine fever attenuated vaccine (Defa *et al.*, 1999) or injected with ovalbumin (Wang *et al.*, 2006). These proteins contain more than 10% threonine and part of the requirement of PDV could be explained by the synthesis of IgA (J.P. Lallès, personal communication). Plasma threonine is after cysteine the second most depressed amino acid in septic rats (Breuillé *et al.*, 1998). However this was shown to affect mainly mucin synthesis in the gastrointestinal tract of rats (Faure *et al.*, 2007). These glycoproteins containing 15-24% of threonine (Lien *et al.* 1997; Piel *et al.* 2004) provide the most popular hypothesis to explain the PDV requirement. In 10-day old piglets, severe chronic deficiency, i.e. 80% of requirement, was shown to induce a diarrhoeic state associated with a decrease in the number of acid mucin producing goblet cells in the duodenum and proximal colon of piglets (Law *et al.*, 2007). Mucin mass and jejunal and ileal villus heights as well as mucosal mass per length unit of small intestine were profoundly affected. In 21-d old piglets fed a threonine-deficient diet at

50% of requirements mucosal protein synthesis was significantly depressed and this was associated with a significant decrease in mucin synthesis rates (Wang *et al.*, 2007).

Surprisingly, however there was a decrease in both mucin and total protein synthesis rates at higher threonine supply (Wang *et al.*, 2007). In piglets of similar age fed an adequate level we also had found a decrease in protein synthesis rate in the intestine as well as in the muscle compared to piglets receiving 70-75% of their requirements (Sève and Ponter, 1997). Recently, in 21-d old piglets of different genotype, we found no significant effect of the same level of chronic dietary deficiency, neither on muscle nor on the intestine, and there was no effect either on mucin synthesis rate (Hamard, 2008). The response of protein synthesis and, presumably, of mucin synthesis to threonine supply could be curvilinear, with a maximum in the range 70-90%. This response would be similar to that of feed intake which was also maximised at this level (Hamard, 2008) and previously shown to decrease in pigs both at lower and at adequate or excess level (Henry and Sève, 1993). Investigating the underlying mechanisms through transcriptomic studies, Hamard *et al.* (2007) found at ileal level that a moderate threonine deficiency repressed two genes assumed to be implied in the down regulation of protein synthesis by individual amino acid starvation, GCN2 coding for a phosphorylase which inactivates the eukaryote initiation factor 2 and the gene coding for 4E-BP1 which inactivates the eukaryote initiation factor 4. Interestingly, GCN2 was shown to be involved also in the control of feed intake when individual amino acid supplies are suppressed. Altogether, this provides further evidence of adaptive mechanisms to amino acid deficiency, here preserving intestine growth from moderate threonine deficiency, as already shown with protein-deficient diets (see above).

Back to the implication of threonine in gut health, there seems to be also a limited risk of depressing mucin production with marginal dietary deficiency. Nevertheless, Hamard (2008) got evidence of mucosa alteration at ileal level, with a decrease in villus height and microvillus-borne aminopeptidase activity and, more importantly, an increase in paracellular permeability with the above-reminded risk of uncontrolled antigen passage or bacterial translocation (Table 5). In parallel an overexpression of tight junction protein genes (ZO-1 and cingulin) was observed, probably as a compensatory regulating mechanism similar to the one assumed for mucosal protein synthesis preservation (Hamard *et al.*, 2007), although apparently less efficient. In addition, there was an increase in ileal mucosa glucose absorption capacity, associated with an overexpression of genes involved in the energy metabolism. This would mean that regardless

Table 5. Impact of moderate threonine deficiency on structure and physiology of the ileum mucosa in 3-wk old weaned piglets (Hamard et al., 2006, 2007; Hamard, 2008).

Diet	Control	Low threonine (70% of control)	Significance (P)
Mucosal architecture			
Villus height, μm	714	446	<0.05
Villus:crypt ratio	3.64	2.46	<0.05
Crypt width, μm	58	53	<0.05
Aminopeptidase N, IU/mg protein	374	269	<0.05
Permeability [1]			
Transepithelial resistance, ohm/ cm^2	50	35	>0.05
FITC dextran 4000 flux, ng/ cm^2/h	1463	2568	<0.05
Glucose inducedchange in short-circuit current, μA/ cm^2 [1]	69	90	0.10

[1] Measured in Ussing chambers.

of their final efficiency, the adaptive mechanisms could increase the energy demand of the gastrointestinal tract. At this stage, we could conclude that a number of adaptive mechanisms aim at preservation of the intestinal mucosa development, but that functionality may be affected in the first place, before tissue growth, by a moderate threonine deficiency.

The implication of both threonine and fibre in the maintenance of gut health has consequences in the practice of diet formulation and technology. In the same way as immune challenges, fibre also induces ileal endogenous losses (see section 2), partly in the form of mucins, and does create a moderate threonine deficiency. Although growth of the gastrointestinal tract is preserved, this is at the expense of carcass body compartments. Before carcass growth retardation, its amino acid composition is affected in the first place whereas gastrointestinal tissues are preserved also at this level (Hamard, 2008). At this point the question is: 'On which bases and how should we cover this additional threonine requirement'?

Taking advantage of variations between diets based on pea from different cultivars (Hess *et al.*, 2000) or induced by technological treatment of a sunflower meal-based diet (Lahaye *et al.*, 2004) nitrogen balance was shown to decrease linearly according to the increase in ileal endogenous losses, as measured through isotope labelling of the diet or the animal (Figure 5). This was obtained in ileorectostomised pigs, without a functional large intestine, therefore it was possible from N data to estimate directly the impact of these losses on threonine balance after checking it was the first-limiting amino acid (Lahaye *et al.*, 2004). This way we could show a decrease in threonine efficiency related to endogenous losses presumably induced by sunflower meal fibre and reduced by pelleting or high-flow extrusion treatment. According to the model described by Sève and Hess (2000), digested dietary nitrogen can be partitioned into that used for ileal endogenous protein and that used for other body protein synthesis. It is then possible to calculate the cost of nitrogen and amino acids secreted in the small intestine. For threonine the total cost was found to be 1.82 times the ileal loss, Thr_e, which means that the metabolic cost was 0.82 times this loss. This corresponded to an efficiency of true digestible threonine for Thr_e K_e = 0.55, whereas the efficiency for other body protein synthesis Thr_b was K_b = 0.74. That efficiency depends on intestinal threonine metabolism parameters that are relatively well known, at least in the upper gut, although their precise measurements are difficult. Firstly, in normal enteral feeding, lumenal threonine is the preferred source for mucosal end mucin synthesis (Nichols and Bertolo, 2008), although, due to the high metabolic plasticity and adaptive capacity of digestive tissues parenteral supplements also can be efficiently used (Law *et al.*, 2007). The first pass extraction of dietary threonine by the PDV provides a measurement of this preference. It represents 40 to 70% of total extraction in

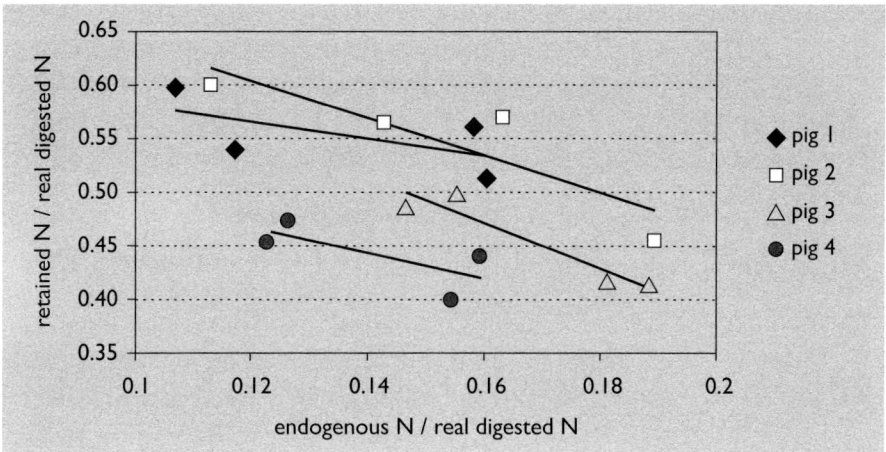

Figure 5. Impact of endogenous nitrogen losses on body N retention (Lahaye et al.*, 2004).*

piglets receiving a high or low protein diet respectively (Schaart *et al.*, 2005) and 58% of total extraction in 30 kg-pigs (Le Floc'h and Sève, 2005). Secondly, the recycling of threonine was non-measurable in piglets (Van der Schoor, 2002; Schaart *et al.*, 2005) and quite low in 30 kg-pigs, representing only 30% of total extraction (Le Floc'h and Sève, 2005). Because, as shown in the latter paper, the irreversible loss of threonine through oxidation is minimal in the PDV and low recycling minimises the flux in the body metabolic pool, the efficiency of absorbed threonine for the synthesis of protein secreted in the small intestine, namely mucins, should be higher than the efficiency of lysine for which both lower first-pass extraction and higher recycling [45% of intake and 85% of total extraction, respectively according to Van der Schoor *et al.* (2002)] were measured (Table 6). Accordingly, in our model of pigs deprived of functional large intestine, the metabolic cost of lysine secreted in the small intestine was higher than that of threonine, 1.78 vs. 0.82 times the ileal loss (Lahaye *et al.*, 2004), the difference in total losses of the two amino acids being only due to the higher threonine content of secreted protein, compared to lysine.

In intact pigs, with the enrichment of the diet in soluble fibre (pectin) a significant decrease in the efficiency of threonine utilisation for growth could

Table 6. Compared threonine and lysine metabolism in pig portal-drained viscera.

Input	Threonine[1,2] Le Floc'h and Sève, 2005		Lysine[1,3] Van der Schoor *et al.*, 2002	
	From diet	From arterial blood	From diet	From arterial blood
	121	315	518	4789
Extraction (% of input)	63 (52%)	46 (14%)	236 (45%)	316 (7%)
Available (% of input)	58 (48%)	269 (86%)	282 (55%)	4473 (93%)
Portal output	360		5193	
Net balance (% of diet input)	45 (37%)		404 (78%)	
Recycling (% total extraction)	33 (30%)		438 (85%)	

[1] Values in μmol/kg/h.
[2] Continously-fed 30-kg pigs.
[3] 12 h fed-12 h fasted 8.8 kg piglets. Values are for the 12 h feeding period.

also be measured (Zhu *et al.*, 2005). Since a similar result was obtained when infusing pectin into the caecum (Zhu *et al.*, 2003), this appeared to be, at least partly, the consequences of events occurring in the lower gut rather than in the small intestine. This was corroborated by the increase in colonic mucosa fractional protein synthesis rates under the same conditions (Figure 6; Libao-Mercado *et al.*, 2007) and was thought to be associated with an increase in colonic endogenous protein secretion and its metabolic cost, in response to bacterial proliferation. Unfortunately, only threonine ileal losses have been measured and no reliable figure is presently available on endogenous threonine loss in the colon due to microbial breakdown of amino acids. A priori, because there is no first-pass extraction, the efficiency of dietary threonine for colonic endogenous secretion should be low, but there is no recycling either and the lower gut metabolism is still a black box, not only for threonine but also for other amino acids. Finally, the evidence of a metabolic cost of total endogenous losses, in the lower as well as in the upper gut, clearly question the reliability of ileal digestibility as an approach of amino acid availability for growth, particularly with high-fibre diets that can be recommended for improved gut and general health.

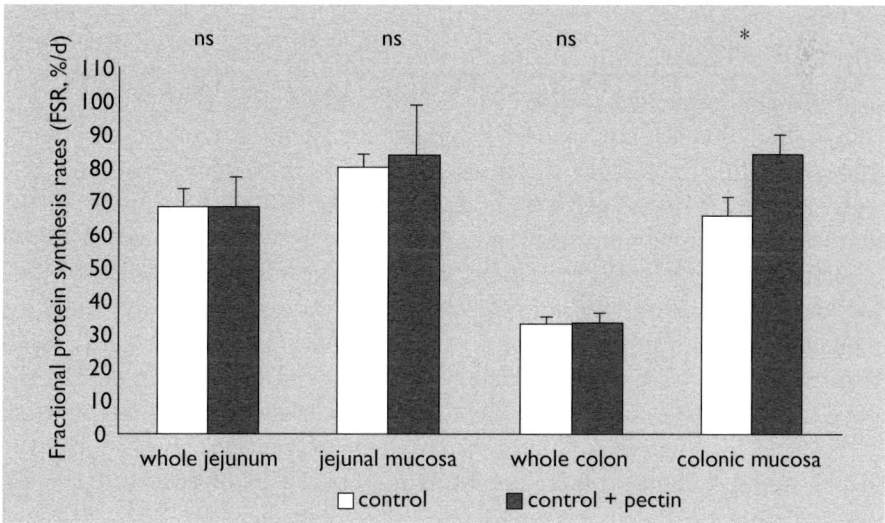

Figure 6. Impact of the enrichment of the piglet diet in pectin on protein synthesis rates in the intestine (Libao-Mercado et al.*, 2007). Significance of dietary treatment effect was tested at* P<0.05.

5. Conclusions

The present literature emphasises the interest of high fibre diets for the prevention of digestive disturbances and infections associated with piglet early-weaning. Due to the immaturity of the digestive capacity, this is particularly true when we use high protein diets with conventional protein sources of low digestibility in order to preserve long-term as well as short-term expression of the growth potential. Fibre plays an important role in balancing microbial fermentation in the ileum and the lower gut. The actual evidence is that the best fibre is made of mixed soluble and insoluble components. However, fibre also induces endogenous losses which have been shown to increase the requirements in some amino acids, namely threonine. In the near-future, it will be necessary to re-consider the availability of amino acids, particularly threonine, from usual protein sources in high-fibre diet. Indeed, the approach of amino acid availability for growth through ileal digestibility measurements assuming additivity of supplies from dietary sources has been questioned.

We have also emphasised the importance of glutamine and arginine, two conditionally essential amino acids, for gut health preservation. These amino-acids play a role in the development and maturation of the digestive system particularly at the mucosa level. They have important physiological and immuno-modulating properties.

However, other nutrients than those discussed in this paper also play a role in gut metabolism and health. For example Koopmans *et al.* (2006) have shown that dietary tryptophan supplements above known requirements increased villi height in the distal small intestine, although we were not able to measure consistent increases in fractional protein synthesis rates in the duodenal and jejunal mucosa (Ponter *et al.*, 1994). Indeed, despite the absence of IDO activity, the main metabolic pathway involved in inflammatory state (Melchior *et al.*, 2004), serotonin synthesis and storage are important in the metabolism of intestinal mucosa and further work is needed on this topic. Similarly, more information is needed on sulphur amino acid metabolism, the PDV extraction of which is also high, namely the role played by the intestine in possibly incomplete transsulfuration (hyperhomocysteinaemia) and the local synthesis of taurine and glutathione respectively involved in cell osmotic balance and in the mucosal antioxidant system (Shoveller *et al.*, 2005). Perhaps more important is the lack of knowledge about the impact of lipids on gut metabolism, functionality and health. We have previously shown that lipids from maize oil, compared to maltodextrin at same energy supply significantly increased protein synthesis rates in the duodenal and jejunal mucosa of early-weaned piglets (Ponter *et al.*, 1994).

References

Adegoke, O.A., McBurney, M.I., Samuels, S.E. and Baracos, V.E. (2003). Modulation of intestinal protein synthesis and protease mRNA by luminal and systemic nutrients. *American Journal of Physiology* **284**: G1017-1026.

Amin, H.J., Zamora, S.A., McMillan, D.D., Fick, G.H., Butzner, J.D., Parsons, H.G. and Scott, R.B. (2002). Arginine supplementation prevents necrotizing enterocolitis in the premature infant. *Journal of Pedatrics* **140**: 425-431.

Aumaitre, A. (1969). Nutritive value of cassava meal and different cereal grans in diets for early-weaned piglets: Digestive utilization and growth performance (in French). *Annales de Zootechnie* **18**: 385-398.

Bach Knudsen, K.E., Laerke, H.N. and Hedemann, M.S. (2008). The role of fibre in piglet gut health. In: *Gut efficiency; the key ingredient in pig and poultry production. Elevating animal performance and health* (Eds. Taylor-Pickard, J.A. and Spring, P.). Wageningen Academic Publishers, the Netherlands, pp. 65-95.

Ball, R.O., Atkinson, J.L. and Bayley, H.S. (1986). Proline as an essential amino acid for the young pig. *British Journal of Nutrition* **55**: 659-668.

Bertolo, R.F., Pencharz, P.B. and Ball, R.O. (1999). A comparison of parenteral and enteral feeding in neonatal piglets including an assessment of the utilization of a glutamine-rich pediatric elemental diet. *Journal of Parenteral and Enteral Nutrition* **23**: 47-55.

Bikker, P., Dirkzwager, A., Fledderus, J., Trevisi, P., le Huërou-Luron, I., Lallès, J.P. and Awati, A. (2006). The effect of dietary protein and fermentable carbohydrates levels on growth performance and intestinal characteristics in newly weaned piglets. *Journal of Animal Science* **84**: 3337-3345.

Boudry, G., Lallès, J.P., Malbert, C.H., Bobillier, E. and Sève, B. (2002). Diet-related adaptation of the small intestine at weaning in pigs is functional rather than structural. *Journal of Pediatric Gastroenterology and Nutrition* **34**: 180-187.

Boudry, G., Guérin, S. and Malbert, C.H. (2004a). Effect of an abrupt switch from a milk-based to a fiber-based diet on gastric emptying in pigs: difference between origins of fibers. *British Journal of Nutrition* **92**: 913-920.

Boudry, G., Péron, V., Le Huërou-Luron, I, Lallès, J.P. and Sève, B. (2004b). Weaning induces both transient and long-lasting modifications of absorptive, secretory and barrier properties of piglet intestine. *Journal of Nutrition* **134**: 2256-2262.

Breuillé, D., Béchereau, F., Buffière, C., Denis, P., Pouyet, C. and Obled, C. (1998). Beneficial effect of amino acid supplementation, especially cysteine, on body nitrogen economy in septic rats. *Clinical Nutrition* **25**: 634-642.

Bruins, M.J., Luiking, Y.C., Soeters, P.B., Lamers, S.W.H., Akkermans, L.M.A. and Deutz, N.E.P. (2004). Effects of long-term intravenous and intragastric L-arginine intervention on jejunal motility and visceral nitric oxide production in the hyperdynamic compensated endotoxaemic pig. *Neurogastroenterology and Motility* **16:** 819-828.

Carey, H.V., Hayden, U.L. and Tucker, K.E. (1994) Fasting alters basal and stimulated ion transport in piglet jejunum. *American Journal of Physiology* **267:** R156-R163.

Coeffier, M., Claessens, S., Hecketsweiler, B., Lavoinne, A., Ducrotte, P. and Dechelotte, P. (2003). Enteral glutamine stimulates protein synthesis and decreases ubiquitin mRNA level in human gut mucosa. *American Journal of Physiology* **285:** G266-273.

Coeffier, M., Miralles-Barrachina, O., Le Pessot, F., Lalaude, O., Daveau, M., Lavoinne, A., Lerebours, E. and Dechelotte, P. (2001). Influence of glutamine on cytokine production by human gut in vitro. *Cytokine* **13:** 148-154.

Defa, L., Changting, X., Shiyan, Q., Jinhui, Z., Johnson, E.W. and Thacker, P.A. (1999). Effects of threonine on performance, plasma parameters and immune function in growing pigs. *Animal Feed Science and Technology* **78:** 179-188.

De Lange, C.F.M., Sauer, W.C., Mosenthin, R. and Souffrant, W.B. (1989). The effect of feeding different protein-free diets on the recovery and amino acid composition of endogenous protein collected from the distal ileum and fecesin pigs. *Journal of Animal Science* **67:** 746-754.

DeMarco, V., Dyess, K., Strauss, D., West, C.M. and Neu, J. (1999). Inhibition of glutamine synthetase decreases proliferation of cultured rat epithelial cells. *Journal of Nutrition* **129:** 57-62.

Domenighi, C., Di Giancamillo, A., Bosi, G. and Arrighi, S. (2006). Can neutraceuticals affect the structure of intestinal mucosa? Qualitative and quantitative microanatomy in L-glutamine diet-supplemented weaned piglets. *Veterinary Research Communications* **30:** 331-342.

Dudley, M.A., Wykes, L.J., Dudley., A.W. Jr., Fiorotto, M., Burrin, D.G., Rosenberger, J., Jahoor, F. and Reeds, P.J. (1997). Lactase phlorizin hydrolase synthesis is decreased in protein-malnourished pigs. *Journal of Nutrition* **127:** 687-693.

Dudley, M.A., Schocknecht, P., Dudley, A.W. Jr., Jiang, L., Ferraris, R.P., Rosenberger, J.N., Henry, J.F. and Reeds, P.J. (2001). Lactase synthesis is pretranslatioally regulated in protein-deficient pigs fed a protein-sufficient diet. *American Journal of Physiology* **280:** G621-628.

Dudley, M.A., Wykes, L.J., Dudley., A.W. Jr., Burrin, D.G., Nichols, B.L., Rosenberger, J.N, Jahoor, F., Heird, W.C. and Reeds, P.J. (1998). Parenteral nutrition selectively decreases protein synthesis in the small intestine. *American Journal of Physiology* **274:** G131-137.

Dugan, M.E. and McBurney, M.I. (1995). Luminal glutamine perfusion alters endotoxin-related changes in ileal permeability of the piglet. *Journal of Parenteral and Enteral Nutrition* **19**: 83-87.

Ebner, S., Schocknecht, P., Reeds, P.J. and Burrin, D.G. (1994). Growth and metabolism of gastrointestinal and skeletal muscle tissues in protein-malnourished neonatal pigs. *American Journal of Physiology* **266**: R1736-1743.

Ewtushick, A.L., Bertolo, R.F.P. and Ball, R.O. (2000). Intestinal development of early-weaned piglets receiving diets supplemented with selected amino acids or polyamines. *Canadian Journal of Animal Science* **90**: 653-662.

Faure, M. Choné, F., Merttraux, C., Godin, J.P., Béchereau, F., Vuichoud, J., Papet, I., Breuillé, D. and Obled, C. (2007). Threonine utilization for synthesis of acute phase proteins, intestinal proteins and mucins is increased during sepsis in rats. *Journal of Nutrition* **137**: 1802–1807.

Hamard, A. (2008). Impact d'une baisse de l'apport de thréonine alimentaire sur la physiologie et le métabolisme de l'intestin grêle du porcelet. PhD. Thesis, National ENSAR–Agrocampus Rennes, France, 140 pp.

Hamard, A., Mazurais, D., Boudry, G., Le Huërou-Luron, I, Sève, B. and Le Floc'h, N. (2007). Physiological aspects and ileal gene expression profile in early-weaned piglets fed a low threonine diet. *Livestock Science*, **108**: 17-19.

Henry, Y. and Sève, B. (1993). Feed intake and ditary amino acid balance in growing pigs with special reference to lysine, tryptophan and threonine. *Pigs News and Information* **14**: 35N-43N.

Hess, V., Sève, B., Langer, S. and Duc G. (2000). Impact of the ileal endogenous protein losses on body nitrogen retention: Towards a new protein evaluation system (in French). *Journées de la Recherche Porcine en France* **30**: 223-229.

Hess, V., Thibault, J.N., Duc, G., Melcion, J.P., Van Eys, J. and Sève, B. (1998). Influence of variety and micro-grinding on pea nitrogen and amino acids ileal digestibility: Real nitrogen digestibility and specific endogenous losses (in French). *Journées de la Recherche Porcine en France* **30**: 223-229.

Hill, J.E., Hemmingsen, S.M., Goldade, B.G. and Dumonceaux, T.J. (2005). Comparison of ileum microflora of pigs fed corn-, wheat-, or barley-based diets by chaperonin-60 sequencing and quantitative PCR. *Applied and Environmental Microbiology* **71**: 867–875.

Hopwood, D.E., Pethick, D.W., Pluske, J.R. and Hampson, D.J. (2004). Addition of pearl barley to a rice-based diet for newly weaned piglets increases the viscosity of the intestinal contents, reduces starch digestibility and exacerbates post-weaning colibacillosis. *British Journal of Nutrition* **92**: 419-427.

Konstantinov, S.R., Ajay Awati, A., Smidt, H., Williams, B.A., Akkermans, A.D.L. and De Vos, W.M. (2004). Specific response of a novel and abundant *Lactobacillus Amylovorus*-like phylotype to dietary prebiotics in the guts of weaning piglets. *Applied and Environmental Microbiology* **70**: 3821-3830.

Koopmans, S.J., Guzik, A.C., Van der Meulen, J., Dekker, R., Kogut, J., Kerr, B.J. and Southern, L.L. (2006). Effects of supplemental L-tryptophan on serotonin, cortisol, intestinal integrity and behaviour in weanling piglets. *Journal of Animal Science* **84**: 963-971.

Lahaye, L., Ganier, P., Thibault, J.N. and Sève, B. (2004). Technological processes of feed manufacturing affect specific protein endogenous losses and amino acid availability for body protein deposition in pigs. *Animal Feed Science and Technology* **113**: 141-156.

Lallès, J.P., Boudry, G., Favier, C. and Sève, B. (2006). High viscosity carboxymethylcellulose reduces carbachol-stimulated intestinal chloride secretion in newly-weaned piglets fed a diet based on skimmed milk powder and maltodextrin. *British Journal of Nutrition* **95**: 488-495.

Larsen, F.M., Moughan, P.J. and Wilson, M.N. (1993). Dietary fiber viscosity and endogenous protein excretion at the terminal ileum of growing rats. *Journal of Nutrition* **123**: 1898-1904.

Law, G.K., Bertolo, R.F., Adjiri-Awere, A., Pencharz, P.B. and Ball, R.O. (2007). Adequate oral threonine is critical for mucin production and gut function in neonatal piglet. *American Journal of Physiology* **292**: G1293-G1301.

Le Bacquer, O., Laboisse, C. and Darmaun, D. (2003). Glutamine preserves protein synthesis and paracellular permeability in Caco-2 cells submitted to « luminal fasting ». *American Journal of Physiology* **285**: G128-136.

Le Bacquer, O., Nazih, H., Blottiere, H., Meynial-Denis, D., Laboisse, C. and Darmaun, D. (2001). Effect of glutamine deprivation on protein synthesis in a model of human enterocytes in culture. *American Journal of Physiology* **281**: G1340-1347.

Le Floc'h, N. and Sève, B. (2005). Catabolism through the threonine dehydrogenase pathway does not account for the high first pass extraction rate of dietary threonine by the portal drained viscera in pigs. *British Journal of Nutrition* **93**: 447-456.

Leterme, P., Froidmont, E., Rossi, F. and Théwis A. (1998). The high water-holding capacity of pea inner fibers affects the ileal flow of endogenous amino acids in pigs. *Journal of Agricultural and Food Chemistry* **46**: 1927-1934.

Li, N., Lewis, P., Samuelson, D., Liboni, K., Neu, J. (2004). Glutaminre regulates Caco-2 cell tight junction proteins. *American Journal of Physiology* **287**: G726-E733.

Libao-Mercado, A.J., Zhu, C.L., Fuller, M.F., Rademacher, M., Sève B. and De Lange, C.F.M. 2007. Intestinal mucosal and whole intestinal protein synthesis in the growing pig: Effect of feeding fermentable fiber. *Livestock Science* **109**: 125-128.

Lien, K.A., Sauer, W.C. and Fenton, M. (1997). Mucin output in ileal digesta of pigs fed a protein-free diet. *Zeitschrift für Ernährungswissenschaft* **36**: 182-190.

Lopez-Pedrosa, J.M., Manzano, M., Baxter, J.H. and Rueda, R. (2007). N-acetyl-L-glutamine, a liquid-stable source of glutamine, partially prevents changes in body weight and on intestinal immunity induced by protein energy malnutrition in pigs. *Digestive Disease Science* **52**: 650-658.

Marchini J.S., Patrick Nguyen P., Deschamps, J.Y., Maugère, P., Krempf, M. and Darmaun D. (1999). Effect of intravenous glutamine on duodenal mucosa protein synthesis in healthy growing dogs. *American Journal of Physiology* **276**: E747-E753.

Mariscal-Landin, G., Sève, B., Colléaux, Y. and Lebreton, Y. (1995). Endogenous amino nitrogen collected from pigs with end-to-end ileorectal anastomosis is affected by the method of estimation and altered by dietary fiber. *Journal of Nutrition* **125**: 136-146.

McDonald, D.E., Pethick, D.W., Pluske, J.R. and Hampson, D.J. (1999). Adverse effects of soluble non-starch polysaccharide (guar gum) on piglet growth and experimental colibacillosis immediately after weaning. *Research in Veterinary Science* **67**: 245-250.

McDonald, D.E., Pethick, Mullan, B.P. and Hampson, D.J. (2001). Increasing viscosity of the intestinal contents alters small intestinal structure and intestinal growth, and stimulates proliferation of enterotoxigenic Escherichia coli in newly-weaned pigs. *British Journal of Nutrition* **86**: 487-498.

Melchior, D., Mézière, N., Sève, B. and Le Floc'h, N. (2005). Is tryptophan catabolism increased under indole amine 2,3 dioxigenase activity during chronic lung inflammation in pigs. *Reproduction Nutrition Development* **45**: 175-183.

Miquel, N., Bach Knudsen, K.E. and Jorgensen, H. (2001). Impact of diets varying in dietary fibre characteristics on gastric emptying in pregnant sows. *Archives of Animal Nutrition* **55**: 121-145.

Nichols, N.L. and Bertolo, R.F. (2008). Luminal threonine concentration acutely affects intestinal mucosal protein and mucin synthesis in piglets. *Journal of Nutrition* **138**: 1298-1303.

Nunez, M.C., Bueno, J.D., Ayudarte, M.V., Almendros, A., Suarez, M.D., Gil, A. (1996). Dietary restriction induces biochemical and morphometric changes in the small intestine of nursing piglets. *Journal of Nutrition* **126**: 933-944.

Papaconstantinou, H.T., Hwang, K.O., Rajaraman, S., Hellmich, M.R., Townsend, C.M., Jr and Ko, T.C. (1998). Glutamine deprivation induces apoptosis in intestinal epithelial cells. *Surgery* **124**: 152-159.

Piel, C., Montagne, L., Sève, B. and Lallès, J.P. (2005). Increasing digesta viscosity using carboxymethylcellulose in weaned piglets stimulates ileal goblet cell numbers and maturation. *Journal of Nutrition* **135**: 86-91.

Ponter, A., Cortamira, N.O., Sève, B., Salter, D.N. and Morgan, L.M. (1994). The effects of energy source and tryptophan on the rate of protein synthesis and on hormone of the entero-insular axis in the piglet. *British Journal of Nutrition* **71**: 66-674.

Piel, C., Montagne, L., Salgado, P. and Lallès, J.P. (2004). Estimation of the ileal output of gastrointestinal glycoprotein in weaned piglet using different methods. *Reproduction Nutrition Development* **44**: 419-435.

Quéméré, P., Bertrand, G. and Chauvel, J. (1977). Comparison of three cereal grains (barley, wheat, maize) in diets for early-weaned pigs (in French). *Journées de la Recherche Porcine en France* **8**: 217-222.

Reeds, P.J., Burrin, D.G., Stoll, B., Jahoor, F., Wykes, L., Henry, J. and Frazer, M.E. (1997). Enteral glutamate is the preferential source for mucosal glutathione synthesis in fed piglets. *American Journal of Physiology* **273**: E408-415.

Rhoads, J.M., Chen, W., Gookin, J., Wu, G.Y., Fu, Q., Blikslager, A.T., Rippe, R.A., Argenzio, R.A., Cance, W.G., Weaver, E.M. and Romer, L.H. (2004). Arginine stimulatesintestinal cell migrationthrough a focal adhesion kinase dependent mechanism. *Gut* **53**: 514-522.

Samuels, S.E., Action, K.S. and Ball, R.O. (1989). Pyrroline-5-carboxylate reductase and proline oxydase activity in the neonatal pig. *Journal of Nutrition* **119**: 1999-2004.

Schaart, M.W., Schierbeek, H., Van der Schoor, S.R.D., Stoll, B., Burrin, D.G., Reeds, P.J. and Van Goudoever, J.B. (2005). Threonine utilization is high in the intestine of piglets. *Journal of Nutrition* **135**: 765-770.

Skiba, F., Callu, P., Lallès, J.P., Thibault,,J. N. and Sève, B. (2007). Digestibilités comparées de l'orge et du blé chez le porcelet en post-sevrage. *Journées de la Recherche Porcine en France* **39**: 153-156.

Sève, B. and Hess, V. (2000). Amino acid digestibility in formulation of diets for pigs: present interest and limitations, future prospects. In: *Recent advances in animal nutrition-2000* (Eds. Garnsworthy, P.C. and Wiseman, J.). Nottingham University Press, Nottingham, UK, pp. 167-181.

Sève, B. and Lahaye, L. (2003). Endogenous protein in the course of digestion: consequences on the availability of amino acids for growth in pigs. In: *Memorias* (Eds. Cuaron, J.A., Cortez, M., Fernandez, D., Labrandero, E., Pérez, V.G. and Quintana, M.) Instituto Nacional de Investigaciones Forestales Agricolas y Pecuarias, Mexico (MEX) *1. Congress of the Latin American Animal Nutrition College, 2003/08/18-23, Cancun (MEX)*, pp. 241-256.

Sève, B. and Ponter, A. (1997). Nutrient-hormone signals regulating muscle protein turnover in pigs. *Proceedings of the Nutrition Society* **56**: 565-580.

Sève, B., Reeds, P.J., Fuller, M.F., Cadenhead, A. and Hay, S.M. (1986). Protein synthesis and retention in some tissues of the young pig as influenced by dietary protein intake after early-weaning. *Reproduction Nutrition Development* **26**: 849-861.

Spanos, C.P., Papaconstantinou, P., Spanos, P., Karamouzis, M., Lekkas, G. and Papaconstantinou, C. (2007). The effect of L-arginine and aprotinin on intestinal ischemia-reperfusion injury. *Journal of Gastrointestinal Surgery* 11: 247-255.

Shoveller, A.K., Stoll, B., Ball, R.O. and Burrin, D.G. (2005). Nutritional and functional importance of intestinal sulfur amino acid metabolism. *Journal of Nutrition* **135:** 609-612.

Stoll, B., Chang, X., Fan, M.Z., Reeds, P.J. and Burrin, D.G. (2000). Enteral nutrient intake level determines intestinal protein synthesis rates in neonatal pigs. *American Journal of Physiology* **279:** G288-294.

Van der Schoor, S.R.D., Reeds, P.J., Stoll, B., Henry, J.F., Rosenberger, J.R., Burrin, D.G. and Van Goudoever, J.B. (2002). The high metabolic cost of a functional gut. *Gastroenterology* **123:** 1931-1940.

Wang, X., Qiao, S.Y., Liu, M. and Ma, Y.X. (2006). Effects of graded levels of true ileal digestible threonine on performance, serum parameters and immune function of 10-25 kg pigs. *Animal Feed Science and Technology* **129:** 264-278.

Wang, X., Qiao, S.Y, Yinn, Y., Yue, L., Wang, Z. and Wu, G. (2007). A deficiency or excess of dietary threonine reduces protein synthesis in jejunum and skeletal muscle of young pigs. *Journal of Nutrition* **137:** 1442–1446.

Wu, G., Davis, P.K., Flynn, N.E., Knabe, D.A. and Davidson, J.T. (1997). Endogenous synthesis of arginine plays an important role in maintaining arginine homeostasis in postweaning growing pigs. *Journal of Nutrition* **127:** 2342-2349.

Wu, G., Flynn, N.E. and Knane, D.A. (2000). Enhanced intestinal synthesis from proline in cortisol-treated piglets. *American Journal of Physiology* **279:** E395-E402.

Wu, G., Knabe, D.A. and Kim, S.W. (2004). Arginine nutrition in neonatal pigs. *Journal of Nutrition* **134:** 2783S-2790S.

Wu, G., Meier, S.A. and Knabe, D.A. (1996). Dietary glutamine supplementation prevents jejunal atrophy in weaned pigs. *Journal of Nutrition* **126:** 2578-2584.

Zhan, Z., Ou, D., Piao, X., Kim, S.W., Liu, W. and Wang, J. (2008). Dietary arginine supplementation affects microvascular development in the small intestine of early-weaned pigs. *Journal of Nutrition* **138:** 1304-1309.

Zhu, C.L., Rademacher, M. and De Lange, C.F.M. (2005). Increasing dietary pectin reduces utilization of digestible threonine intake, but not lysine intake, for body protein deposition in growing pigs. *Journal of Animal Science* **83:** 1044-1053.

Zhu, C.L. (2003). *Effects of graded levels of dietary pectin on amino acid utilization for body protein deposition in growing pigs.* M.Sc. Thesis. Univ. of Guelph, Guelph, Ontario, Canada.

Novel strategies to manage the mycotoxin menace

Alexandros Yiannikouris
Coordinator of Glycomics Research, Alltech North American Center of Animal Nutrigenomics and Applied Animal Nutrition, Alltech Inc., Nicholasville, Kentucky, USA

1. Introduction

Following the crises on bovine spongiform encephalopathy, salmonellosis, listeriosis, the presence of pesticides or chemical compounds such as dioxins in food or feeds, the use of anabolic steroids in farm animals, and the use of GMOs, consumers have become highly sensitive to the notion of food safety. The risks associated with infection and parasites are well understood, but the risks associated with the natural presence of toxins or their metabolites in animal feeds are mostly unknown. Most people consider that natural products are safe. However, contamination of human or animal food with natural toxins may result in a number of problems and even severe diseases. Moulds may grow on plants in the field or during the storage period. These fungi may produce toxins, which may have deleterious effects on humans or animals consuming the contaminated product. Such cases of poisoning may cause death in animals, but are rarely fatal in humans (Pfohl-Leszkowicz, 2000). The history of humanity clearly shows that the mycotoxicological risk has existed since the very beginnings of organised agricultural production (Pittet, 1998). Some references to ergotism in the Old Testament (Schoental, 1984) and fusariotoxins such as T-2 toxin and zearalenone are thought to be responsible for the decline of the Etruscan civilisation (Schoental, 1991) and the Athenian crisis, which occurred in the Fifth Century B.C. (Schoental, 1994). Certain Egyptian tombs are also thought to contain ochratoxin A, responsible for the mysterious deaths of several archaeologists (Pfohl-Leszkowicz, 2000). It was not until 1960 that the first cases of mycotoxicosis were demonstrated following an agricultural incident in Great Britain. More than 100,000 turkeys died of severe liver necrosis and biliary hyperplasia caused by a family of molecules identified as aflatoxins. A large number of mycotoxins have since been discovered, the last major group of which, the fumonisins, was described in 1988. There is also an indirect risk due to the carryover of toxins and their metabolites to edible animal products such as milk, meat, and eggs.

Mycotoxins are secondary metabolites secreted by moulds, mostly belonging to the three genera *Aspergillus, Penicillium* and *Fusarium* produced in cereal grains as well as forages before, during and after harvest. Forages and cereals naturally come into contact with fungal spores. The fungal contamination of plants and the biosynthesis of toxins depend on the state of health of the plant before harvest, meteorological conditions, harvesting techniques, delays and hydrothermal conditions before stabilisation for conservation and feed processing. Depending on the fungus, fungal growth is then controlled by a number of physico-chemical parameters including the amount of free water (aw), temperature, presence of oxygen, nature of the substrate, and pH conditions. Mycotoxins could be then formed pre-harvest as well as post-harvest in storage. Rodents, birds and insects may facilitate contamination by causing physical lesions on plants, providing a route of entry into the plant for fungal spores (CAST, 2003). Thus, mycotoxins could be present throughout food and feeds derived from plants and especially in processed feeds. Avoiding mycotoxin occurrence in the food chain involves a clear understanding of mycotoxin management strategies and effective quality control of food and feeds through adequate sampling, detection and quantification methodology. Nevertheless, it is difficult to estimate precisely the mycotoxin concentration in a large bulk lot because of their uneven distribution in commodities and the large variability associated with the overall mycotoxin isolation procedure. Even when using accepted test procedures, there is variability associated with the selection of the sample, the grinding/mixing/homogenisation procedure and the extraction method used. Interferences could be then encountered with some vegetal matrix giving strong background and contributing to variability by binding and/or masking the real mycotoxin level. Because of this variability, the true mycotoxin concentration in feeds cannot be determined with 100% certainty (Liu *et al.*, 2005). These toxins are thus able to pass through the conventional routine screening methods and then may be released following food-processing or further digestion of the vegetal matrix. It was established that the extraction by solvolysis for further analysis using gas chromatography technique of samples with trifluoroacetic acid resulted in a release of an additional 58% of deoxynivalenol mycotoxin and an increase of 9 to 88% in a set of verification samples compared to results obtained with conventional methods (Zhou *et al.*, 2007). This interaction specifically implicated conjugated or masked mycotoxin through a covalent bonding with more polar molecules such as glucose (Berthiller *et al.*, 2005). Conjugation of mycotoxins to form glycosides and glucuronides is part of the biosynthetic pathway by which plants metabolised/detoxified mycotoxins (Engelhart *et al.*, 1988). Also animals are able to form those conjugates as it was established in swine at the urinary and biliary level (Corley *et al.*, 1985). Transformation would thus represent an alternate metabolic pathway and is not limited to a particular mycotoxin as experiments using liquid-mass spectrometry methodology were able to detect a zearalenopne-4-β-D-glucopyranoside occurring both after digestion in swine of the feed material but also present in naturally infected wheat samples (Schneweis *et al.*, 2002). Furthermore, significant total amounts may be present in feed at low concentrations, below the regulatory levels, able to induce chronic adverse effects that are still poorly understood due to the lack of scientific data in this respect. It may also result in impaired immunity and decreased resistance to infectious diseases (Richard *et al.*, 1978).

Fungal contamination affects both the organoleptic characteristics and the alimentary value of feed, and entails a risk of toxicosis. The biological effects of mycotoxins depend on the ingested amounts, number of toxins, duration of exposure and animal sensitivity. Mycotoxins can induce health issues that are specific to each toxin as described below, have synergistic toxic properties in the frequent case of multi-contamination (Smith *et al.*, 1997), and/or affect the immune status of animals promoting infections and have a negative impact on livestock production. In addition, the possible presence of toxic residues in edible animal products when animals are fed contaminated feeds may have some detrimental effects on human health. The excretion of toxins and their metabolites in milk represents another route by which these compounds may be eliminated from the animal. This process may involve intercellular filtration, passive diffusion across membranes or active transport via secretion vesicles. Aflatoxin B_1 (AFB_1), ochratoxin A (OTA), zearalenone (ZEA) and their metabolites, particularly aflatoxin M_1 (AFM_1), can represent a potential risk to the consumer due to their carryover in cow's milk. AFM_1 was found at the maximum concentration in milk two days after ingestion of AFB_1 by dairy cows (Whitlow *et al.*, 2000; Diaz *et al.*, 2002). The same authors also indicated that AFM_1 disappeared 4 days after AFB_1 was removed from the diet. This means that the mechanisms involved in the mycotoxin transfer do not need adaptation.

In pigs, naturally deoxynivalenol (DON)-contaminated feeds induce clinical effects such as diminished feed consumption and lower weight gain for doses inferior to 2 mg/kg of feed (Rotter *et al.*, 1994; Trenholm *et al.*, 1984). At 1.3 mg/kg of feed, DON has a negative effect on feed intake by growing pigs, with signs of feed refusal. The feed refusal was total when 12 mg/kg were reached and vomiting induced at 20 mg/kg. The most common signs of the presence of DON are growth depression, loss of appetite, vomiting, and lesions of the intestinal tract. The situation with naturally contaminated toxin is by far more complex because of the implication of several different toxins even when present at chronic contamination levels. Other observations accounted for ZEA-contaminated feedstuffs inducing stillbirths, neonatal mortality, foetal mummification, splay-leg of pigs, abortion, abnormal return to oestrus, and other abnormalities (Diekman *et al.*, 1992). OTA leads to important changes in the renal function of pigs with the impairment of proximal tubular function, altered urine excretion, and increased excretion of urine glucose (CAST, 2003). OTA is also able to enhance the susceptibility of pigs to natural infection by *Salmonella cholerasuis*, *Serpulina hyodysenteriae*, or *Campylobacter coli* (Stoev *et al.*, 2000). It has been generally admitted that mycotoxins might thus predispose livestock to infectious disease, possibly resulting in feed refusal and decreased productivity. Mycotoxins by their immunosuppressive properties may facilitate the entrance of pathogens that may overshadow the primary cause of the infection due to the toxin itself.

Three steps to control mycotoxins have been identified (Jouany *et al.*, 2007); (1) A first control step may be used before any fungal infection has been detected. Thus the control of mould growth involves maintaining the physical integrity of cereal grains with the aim of limiting the access of moulds to nutrients present in the grains, and the strict control of environmental conditions. Good agricultural practices could be applied by implementing crop rotations, adequate tillage, use of soil fertilisers, periods of planting, use of fungicides or aromatic plant essential oils, controlling the presence of insects, rodents, weed, but also the possible use of breeding transgenic resistant crops, or installing a microbial competition between strains of fungi using atoxigenic strains. A new approach has been proposed, involving the isolation of *Aspergillus flavus* and *A. parasiticus* strains that do not produce aflatoxins, using them to out-compete natural toxin-producing strains (Cotty *et al.*, 1994). These strains occupy the same ecological niche as toxin-producing strains, so they decrease the level of contamination with toxin-producing moulds. Finally, models could be used with integrated regression to evaluate and predict the risk for a contamination to occur; (2) During the periods of fungal invasion of the plant material, there is a possible mycotoxin production. Thus the timing of harvest and the physiological stage of the plants could be of prime importance as well as the technical parameters used to harvest feeds. Cleaning, prevention of soil contact, humidity level before and during storage and temperature are factors that need to be considered to limit the entrance of the moulds and the synthesis of mycotoxins; (3) Finally, the third step resides in the decrease of the mycotoxin level of the feed after having identified the contamination and if the contamination is at acceptable levels as per the regulatory statements. Physical methods such as grain cleaning by washing with water or sodium carbonate to reduce the contamination of maize with *Fusarium* toxins, manually sorting out contaminated grains by the physical aspect of grains or by fluorescence to detect the presence of some mycotoxins, have been used. Other toxin inactivating techniques have been proposed such as high temperature, UV, X-rays or microwave irradiation, and solvent extraction of toxins (Scott, 1998) could be applied. Drying is thus an essential step in the preservation process of dry feed, and anaerobiosis is a prerequisite to the storage of feed in a moist form. A variety of chemical agents such as acids, bases (ammonia, caustic soda), oxidants (hydrogen peroxide, ozone), reducing agents (bisulphites), chlorinated agents and formaldehyde, have been used to degrade mycotoxins in contaminated feeds, particularly aflatoxins and may provide additional guarantees if there is a predictable risk. Finally biological treatments could be implemented by the use of specific enzyme to detoxify the mycotoxin and generate secondary non-toxic metabolites (which isn't always possible!). This approach uses the ability of certain micro-organisms to metabolise mycotoxins (*Corynebacterium rubrum*) in contaminated feed or to biotransform them (*Rhizopius, Aspergillus, Eurotium*, Nakazato *et al.*, 1990). However, these biological processes are generally slow and have a low efficiency.

However the mycotoxin risk remains unavoidable and the lack of practical solutions to totally avoid mycotoxin contamination of feed explains why inorganic materials such as clays, bentonites and aluminosilicates, known for their adsorptive properties, were first proposed to reduce the toxic effect of aflatoxins (Grant and Phillips, 1998). These inorganic materials (hydrated sodium calcium aluminosilicates (HSCAS), phyllosilicates, zeolites, cholestyramine, polyvinylpolypyrrolidoxynivalenol) showed only limited efficacy against other mycotoxins. Furthermore, they reduced the biological value of certain nutrients and could contain dioxins and heavy metals. They may have adverse nutritional effects due to the large amounts included in the diet and their possible interaction with certain minerals or vitamins. Some of these binders are not biodegradable and may affect the efficacy of retention basins and digesters used for the treatment of animal effluents. Clays render wet floors slippery, increasing the risk of accidents for both animals and human staff. Certain strains of lactic acid bacteria, propionibacteria and bifidobacteria have cell wall structures that can bind mycotoxins (Ahokas *et al.*, 1998) and limit their bioavailability in the animal body. Mycotoxins are then eliminated in the faeces. Organic materials such as yeast cell walls (Devegowda *et al.,* 1998) have been proposed as an alternative solution to complex several mycotoxins within the gastrointestinal tract without impairing nutrient bioavailability or inducing detrimental environmental effects. Their sequestering efficacy results from the large area available for exchange. It is important to understand the chemistry of the sequestration process involved in toxin clearance. More than simple 'binding assays' (Yiannikouris *et al.*, 2003), interaction kinetic models based on overall capacity, standardised affinity rate and stereochemical views are becoming available. These models are required to complete and support in a meaningful way the advances in applied nutrition needed to define the beneficial role of organic sequestrants prepared from yeast cell wall.

2. Materials and methods

2.1. Yeast biomass production

Four strains of *S. cerevisiae* were tested: the wild type wt292, the fks1 mutant type, the mnn9 mutant type, and sc1026 (Alltech, Inc., KY, USA). They were grown in flasks containing an YPD medium (1% (w/v) yeast extract, 2% (w/v) bacteriological peptone and 2% (w/v) glucose) at 30°C, shaken at 200 rpm. The growth of these cultures was stopped at $2x10^7$ cells/ml.

2.2. Isolation of cell walls and cell wall characteristics

Cell walls were disrupted with glass beads and isolated (Dallies *et al.*, 1998). Quantification of mannans and glucans was achieved by 2 N H_2SO_4 hydrolysis at 100 °C for 4 h. An enzymatic method was used to determine the chitin content (Popolo *et al.*, 1997), Mannan/glucan ratios were 1.25, 2.09, 2.18, and 0.21 in the cell walls of wt292, fks1, sc1026, and mnn9, respectively. Wt292 and mnn9 strains had high β-D-glucans contents, i.e. 45 and 75%, respectively, of the total cell wall compared to fks1 and sc1026 strains (~30%). Both mnn9 and fks1 strains showed high chitin contents of 9.7 and 5.8%, respectively, as compared to ~2% for wt292 and sc1026.

2.3. Alkali extraction of carbohydrates from yeast cell wall

S. cerevisiae cell walls were fractionated by alkali extraction with 1 M NaOH and 0.5% of $NaBH_4$ for 24 h at 37 °C under agitation, adapted from Fleet (1991) and Catley (1988). The suspension was centrifuged (10,000 g for 5 min) and the supernatant and pellet fractions were separated. Supernatants were dialyzed (1:100, v/v) on cellulose-ester membranes (MWCO: 6 to 8,000) with 0.02 M Tris/HCl buffer (pH 7.4) for 16 h at 4 °C with magnetic stirring. B-D-glucans, which were separated from mannans using a concanavalin A sepharose column, were eluted with 0.02 M Tris/HCl buffer (pH 7.4), 0.5 M NaCl and stored at -20 °C until use. The pellets were thoroughly washed with 1 ml of 75% ethanol/10 mM Hepes buffer (pH 7.1) and then suspended in 2 ml of 0.1 M Tris/HCl buffer (pH 8.5), and finally stored at -20 °C until use.

2.4. Quantitative analysis of cell wall carbohydrates

Analyses were performed in triplicate on a Dionex Bio-LC system (Dionex, Sunnyvale, CA, USA) with a pulsed amperometric detector equipped with a gold electrode. Separation of carbohydrates was performed on a CarboPac PA1 anion-exchange column (4 × 250 mm) equipped with a guard column. Elution was at a flow rate of 1 ml/min at 20-22 °C with 18 mM NaOH.

2.5. Mycotoxin quantification

Mycotoxins were analysed by isocratic HPLC on a HP-1090 Series II HPLC (Hewlett-Packard Co.) using a UV diode array detector coupled to an HP-1046A fluorescence detector (Ex. = 280 nm; Em. = 460 nm). A C18 and a C8 Nucleosil Spherisorb ODS-2 column (4 × 150 mm) equipped with a guard column was used at a flow rate of 0.8 ml/min.

2.6. In vitro *technique to estimate the mycotoxins complex-forming capacity of yeast cell walls*

For each *in vitro* test, 100 µg/ml of adsorbent was placed in tubes together with 2, 4, 6, 8, 10 or 20 µg/ml of mycotoxin dissolved in water and agitated at 200 rpm during 1.5 h at 37 °C, before centrifugation at 5,000 g. The amount of bound mycotoxin was calculated by subtracting the amount of free toxin found in the supernatant of the experimental tubes from the amount found in control tubes with no adsorbent. DataFit 7.1 software (©Oakdale) was used to plot the experimental data, set up the regression curve (curve fitting) and calculate the statistical data (Yiannikouris *et al.*, 2003).

2.7. Wide-angle X-ray diffraction patterns of β-D-glucans

Laminarin, a water soluble linear (1→3)-β-D-gucan branched with short (1→6)-β-D-glucans side-chains extracted from *Laminaria digitata*; and curdlan, an insoluble linear (1→3)-β-D-glucan extracted from *Alcaligenes faecalis*, were investigated by X-ray in their semi-crystalline form and in their hydrated form (Yiannikouris *et al.*, 2004d). Laminarin and curdlan powders were hydrated over a solution of $BaCl_2$ ($a_w = 0.9$ at 25 °C during 5 to 8 days) until equilibrium of the wet sample mass was reached. X-ray diffraction measurements were performed using an Inel X-ray G 3,000 equipment (Nuessli *et al.*, 2003).

2.8. In silico molecular mechanics investigations

Molecular modeling was carried out on Silicon Graphics computers with Accelrys packages (Accelrys, Inc, San Diego, CA, USA) and InsightII, Biopolymer, Analysis, Docking and Discover modules. A CFF91 force-field adapted to polysaccharide/ligand interaction studies was used in vacuum condition ($\varepsilon=1$). Construction of the most highly probable conformations of (1→3)-β-D-glucan chain was performed as described in previous work (Yiannikouris *et al.*, 2004d). [φ,ψ] and [$\varphi \psi,\omega$] dihedral angles values of (1→3)- and (1→6)-β-bonds between two glucoses were explored through their rotation (from -180° to +180°) and their minimum energy conformations evaluated before further elongation to a 5,868 Da polymer comparable to laminarin molecule.

2.9. Liquid NMR investigations of the interaction between ZEA and pure β-D-glucans

Fifty µg/ml of both molecules were added to 10 m of a Milli-Q^{uf+} water, shaked at 640 rpm during 1.5 h at 39 °C, and then cooled on a bath of 2-propanol kept at -30 °C before freeze-drying. The powder containing the [laminarin + ZEA] complex was solubilised in 500 µl of DMSO-d_6 to reach the concentration of 1 mg/ml for both molecules. NMR measurements were carried out on a Bruker Avance 400 spectrometer (300 K). The ^1H NMR spectra of the [laminarin + ZEA] complex was recorded and compared with the spectra of the two separated components in same conditions (Yiannikouris *et al.*, 2004d).

2.10. In silico molecular mechanics investigations of the interaction between mycotoxins and β-D-glucans

ZEA, AFB_1, DON, PAT molecules were constructed using NMR and X-ray data and data of Panneerselvam *et al.*, 1996 and Cordier *et al.*, 1990 (Figure 1). A [ZEA + β-D-glucans] complex was formed by manually positioning mycotoxins molecules in the cavity offered by the (1→3)-β-D-glucan polymer. Translations plus rotations as well as up and down positioning of mycotoxins were achieved inside the β-D-glucan to find all the possible spatial orientations of the interaction, within a 10,000 iterations minimisation procedure (Yiannikouris *et al.*, 2004d).

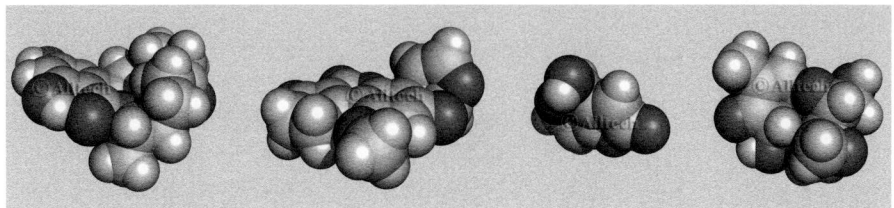

Figure 1. Computer-generated views of the energy-minimized conformations of ZEA, AFB_1, PAT and DON molecules, respectively.

3. Results and discussion

3.1. Complex forming capacity of yeast cell walls

The chemical mechanisms involved in the sequestering activity of *Saccharomyces cerevisiae* cell wall components with several major mycotoxins were investigated. The *in vitro* methodology was based on the comparison of several sources of yeast cell wall differing in their relative glucan/mannan/chitin content. The evaluation of the affinity rate was achieved using Hill's model and sub-models establishing a cooperative interaction between the mycotoxin level contained in a medium and the level of mycotoxin sequestered. Curves representing the amount of mycotoxin bound *vs.* the amount of mycotoxin added were plotted according to Hill's model with n sites (HMN) given by Equation (1) in Table 1, with $R^2 = 0.993$ (Yiannikouris *et al.*, 2003). Equation (1) was used to set up two sub-models: (a) Equation (2) took into account the amounts of total β-D-glucans in the cell wall related to the adsorption properties for ZEA (Yiannikouris *et al.*, 2004a); (b) Equation (3) took into account the respective roles and amounts of alkali-insoluble and alkali-soluble β-D-glucans in the adsorption process (Yiannikouris *et al.*, 2004b). Each equation led to the calculation of the affinity rates A (Equation 4) either for the soluble or insoluble forms of glucans added in the same glucose concentration. A correlation between the amount of β-D-glucans in cell walls and complex-forming efficacy was demonstrated ($R^2 = 0.889$; RSD = 0.534 μg/ml; A ≈ 30%). Cell walls of wt292 and mnn9 strains with higher levels of β-D-glucans were able to complex larger amounts of ZEA with higher affinity rates than the fks1 and sc1026 strains. The higher chitin content in mnn9 and fks1 strains results in higher insolubility of β-D-glucans, less flexibility of the overall structure and, therefore, a decrease in the access of ZEA to the chemical sites of the β-D-glucans. Thus, these strains exhibited a lower complex-forming capacity than expected from their β-D-glucans content (Yiannikouris *et al.*, 2004a).

Table 1. Models used to plot and evaluate ZEA adsorption on S. cerevisiae cell wall β-D-glucans according to Hill's equation.

Model expression	$T_{bound} = f(T_{added})$
Hill's model with n sites (1)	$T_{bound} = [T_{bound}^{max} \cdot (T_{added})^n] / [KD + (T_{added})^n]$
Hill's model with n sites taking into account the amount of β-D-glucans (2)	$T_{bound} = (Glucans \times T_{bound}^{max} \cdot (T_{added})^n) / ((Glucans \times KD) + (T_{added})^n)$
Hill's model with n sites taking into account the amount of alkali-soluble and alkali-insoluble β-D-glucans (3)	$T_{bound} = \alpha \cdot [T_{bound}^{max} \cdot (T_{added})^n] / [\alpha \cdot KD + (T_{added})^n]$ and $\alpha = (S_{ol} + a \cdot I_{nsol}) \cdot \beta\text{-}D\text{-}glucans$
Affinity rate, in percent (4)	$A = T_{bound}^{max} / 2 \cdot \sqrt[n]{KD}$

Note: a = variable distinguishing the role played by alkali-insoluble compared with alkali-soluble β-D-glucans; I_{nsol} = amount of alkali-insoluble β-D-glucans (%); S_{ol} = amount of alkali-soluble β-D-glucans (%); KD = association constant compared with amount of toxins initially added; T_{bound} = amount of bound toxin (μg/ml); T_{added} = amount of toxin added (μg/ml); T_{bound}^{max} = maximal amount of bound toxin (μg/ml).

The alkali-insoluble fraction had a greater affinity for ZEA adsorption (up to 50%) than the alkali-soluble fraction (about 16%). From the results obtained with the individual fractions and plotting our data with the Equation (3) (R^2 = 0.969; RSD = 0.296 μg/ml), it was confirmed that mnn9 and wt292 strains (A = 50.4 and 35.9% for the alkali-insoluble fractions, respectively) had the highest ZEA adsorption efficacy. Elimination of a part of chitin content during alkali extraction of yeast cell wall improved 1.5 times the capacity of the alkali-insoluble fraction of β-D-glucans of mnn9 strain to adsorb ZEA as compared to wt292 (Yiannikouris *et al.*, 2004b).

A correlation between the amount of β-D-glucans in cell walls and complex-forming efficacy was also shown for AFB1, DON, PAT and, to a lesser extent, for OTA (R^2 values between 0.980 and 1.000 and RSD below 0.855 μg/ml) (Table 2). For all of these mycotoxins, cell walls of *wt292* and *mnn9* yeast strains with higher levels of β-D-glucans are able to complex larger amounts of toxins with higher affinity rates than the *fks1* and *sc1026* strains. Again, the higher chitin content in *mnn9* and *fks1* strains results in higher insolubility of β-D-glucans, less flexibility of the overall structure and, therefore, a decrease in

Table 2. Evaluation of the biological parameters for the sequestration of AFB_1, PAT, DON and OTA using Hill's equation integrating the amount of soluble and insoluble fractions of β-D-glucans extracted from four strains of S. cerevisiae.

	Strains	AFB_1	PAT	DON	OTA	ZEA
Affinity (%)	wt292	95.0	26.7	94.8	13.6	35.9
	fks1	37.7	26.4	50.5	12.2	17.4
	sc1026	45.0	26.9	64.4	12.4	22.2
	mnn9	72.5	37.0	84.0	13.9	50.4
R^2 (HMN)	wt292	1.000	0.995	0.999	0.989	0.948
	fks1	0.995	0.995	0.999	0.994	0.980
	sc1026	1.000	0.999	0.998	0.980	0.955
	mnn9	0.999	0.994	0.999	0.997	0.991

Note: R^2 = coefficient of determination; RSD = residual standard deviation; HMN = Hill's model with n sites.

the access of mycotoxins to the chemical sites of the β-D-glucans. Thus, these strains exhibit a lower complex-forming capacity than expected from their β-D-glucan contents (Yiannikouris *et al.*, 2004a). Differences between mycotoxins were also recorded and could be undoubtedly assigned to the stereochemical, electrical charge, solubility, nature and size discrepancies between mycotoxins (Figure 1). Their affinities for the 3D-structure of the binding sites offered by β-D-glucans are consequently variable and ranked from 95.0 to 12.2% of affinity as follows: AFB_1 > DON > ZEA > PAT > OTA. Another study showed a significant efficacy for sequestering T-2 toxin, as well as endophyte associated toxins. The extraction of the glucan fraction from the entire yeast cell wall contributed to increasing the potential of glucans to sequester mycotoxins. Differences were then characterised in relation to the structure of the glucan fraction extracted define by the differences observed between the insoluble fraction and the soluble fraction of glucans. The structure of glucans available in *mnn9* and *wt292* strains had the highest ZEA sequestration efficacy (R^2 = 0.969; RSD = 0.296 µg/ml). Elimination of a part of the chitin content during extraction of yeast cell wall improved the sequestrant activity by 150% (Yiannikouris *et al.*, 2004b).

3.2. Mechanism of complex forming

The affirmation of the chemical interaction between mycotoxins and sequestrant was demonstrated using molecular mechanics investigation, because the affinity was not only dependent on the quantities of β-D-glucans present in the cell wall but also on their structure and the network organisation defined in their single and/or triple helix structure. β-D-glucans were defined as complex three-dimensional structures with alternating regions of random coil, single helices and triple helices (Kogan, 2000). Merged with our previous results and according to investigations carried out on the structure of yeast cell wall, the alkali-insolubility of β-D-glucans have a direct role in maintaining rigidity and integrity of the yeast cell wall. In contrast to the alkali-insoluble glucans, alkali-soluble β-D-glucans might make the yeast wall more flexible (Fleet, 1991). From the wide-angle X-ray diffraction investigation, curdlan spatial organisation was calculated and evaluated as a cluster of four triple-helix chains of (1→3)-β-D-glucan (Chuah *et al.*, 1983). These helical chains were 1.56 nm apart, with a fiber period of 0.60 nm, consequently defining six β-D-glucopyranose units per turn of helix. Such organisation was assessed by the measurement of the 2Q diffraction angles and reticular distances (d_i) and compared to the well known paramylon extracted from *Poria cocos*, an other water insoluble (1→3)-β-D-glucan molecule, that had a lower d_i maximal value than the d_i maximal value obtained with curdlan, respectively 1.36 (Chuah *et al.*, 1983) and 1.50 nm. Consequently, it was proposed that curdlan would involve a more relaxed triple helical conformation as compared to paramylon (Marchessault and Deslandes, 1979). In this respect, since a higher reticular distance of 1.62 nm was obtained for laminarin at both hydrated levels, we could hypothesise that laminarin is organised as a highly relaxed triple-helix and/or single-helix. Using molecular mechanics, we were then able to perform the investigation of the most stable conformation of a short chain of β-D-glucans using the above structural data obtained with X-ray diffraction. Our work calculating the potential energy of the glucan structure, confirmed that the most stable conformation of the (1→3)-β-D-glucan single helical chain was found for a six β-D-glucopyranose units per turn with a glycosidic linkage of $[\varphi,\psi] = (-100°, +140°)$. These investigations applied to model glucan molecules were the only way to record structural data which could be used to approach the glucan organisation of the yeast cell wall.

Complementary nuclear magnetic resonance (NMR) and X-ray structural data analysis (Chuah *et al.*, 1983), was used to assess the overall stability of the modelled molecules in their most stable possible conformations and to allow calculation of the statistical probability of the existence of each conformation. These helical chains were defined by six β-D-glucopyranose units per turn of helix. Macromolecular *in vitro* experiments and NMR experiments (Yiannikouris *et al.*, 2003; 2004a-d) demonstrated that environmental conditions such as pH and nature of the solvent, can decrease the stability of the geometry of β-D-glucans, and thus decrease their ability to complex mycotoxins. They had the potential to relax their structure and form single helices, depending on surrounding conditions, leading to dynamic behaviour of the β-D-glucans. Thus, environmental conditions such as pH and nature of the solvent can decrease the stability and alter the geometry of β-D-glucans (Yiannikouris *et al.*, 2004c).

Soluble ^{1}H NMR investigations were achieved using soluble glucan, which was taken as a model of single helix glucans comparable to glucans found in yeast cell wall, as proven in X-ray diffraction experiments. NMR spectrum analysis showed that the interaction of ZEA and soluble glucan strongly reduced the two peaks of hydroxyl groups of the ZEA phenol moiety (δ = 10.07 and δ = 10.86 ppm), which thus means that these hydroxyl groups are involved into hydrogen bonds in the [ZEA-glucan] complex. Molecular mechanics showed that the (1→3)-β-D-glucan chain favoured a very stable intra-helical association with ZEA (Figure 2). Two types of bonds were identified in monitoring the association: (1) ZEA hydroxyl, ketone and lactone groups were involved in hydrogen bonds with (1→3)-β-D-glucan, and (2) a stabilising van der Waals electronic interaction occurred due to the geometrical symmetry and to the proximity between β-D-glucopyranose rings and the ZEA phenol moiety. Therefore, the ZEA molecule is totally trapped inside the β-D-glucan single helical structure (Yiannikouris *et al.*, 2004d).

The site-specific interactions between a mycotoxin and the sequestrant were then investigated by means of *in silico* molecular modelling. ZEA, AFB_1, DON and PAT molecules were constructed using NMR and X-ray data (Panneerselvam *et al.*, 1996; Cordier *et al.*, 1990; Marchessault and Deslandes, 1979). A [ZEA + β-D-glucan] complex was formed by manually positioning mycotoxin molecules in all possible positions in the cavity offered by the (1→3)-β-D-glucan polymer (Yiannikouris *et al.,* 2004d). It was concluded that the defining chemical interactions involve weak chemical linkages such as hydrogen bonds and van der Waals forces responsible for a stacking effect occurring between β-D-glucans and the hydroxyl and cyclic groups of mycotoxins, if available. The consequence was the production of several *in silico* models showing mycotoxin molecules caged inside the helix-shaped

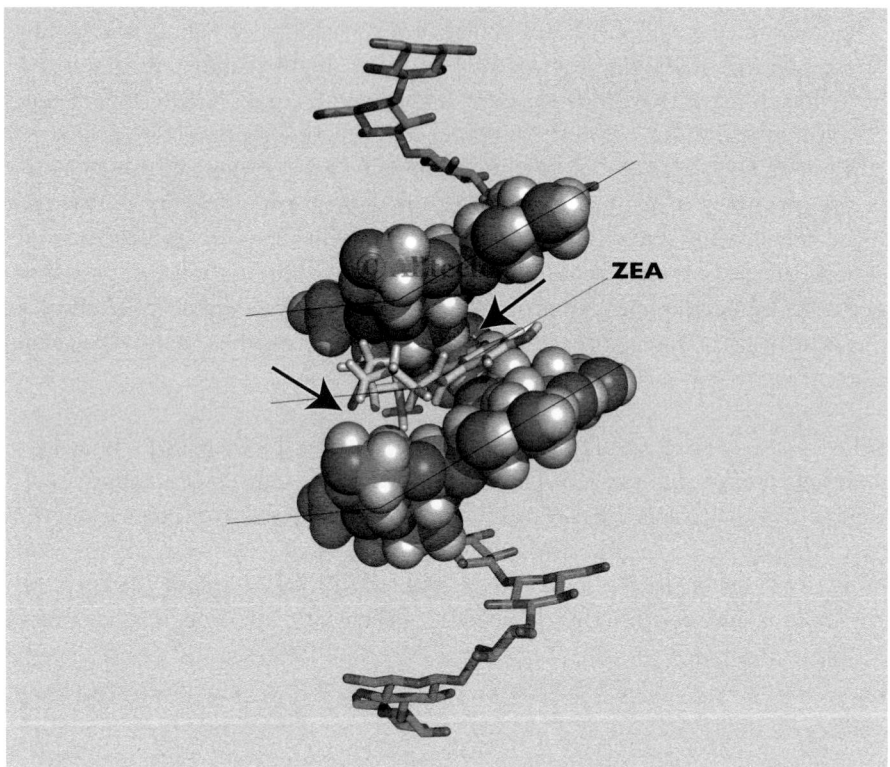

Figure 2. Computer-generated views of the docking of ZEA into the single-helix of (1→3)-β-D-glucan chain. Arrows indicate hydrogen bonds involved in the interaction. Lines highlights the steric complementarities between ZEA and (1→3)-β-D-glucan geometry.

(1→3)-β-D-glucan, which were firmly stabilised by (1→6)-β-D-glucan branched side chains. ^{1}H NMR analysis showed that the interaction of ZEA and β-D-glucans strongly implicated hydroxyl groups of the ZEA phenol moiety. This approach showed that a key feature of the glucan molecule was the geometric similarity between the spatial organisation of mycotoxin molecules and the active site on the single-helix conformation containing six β-D-glucopyranose residues per turn of the (1→3)-β-D-glucan chain (Figure 2). Using molecular modelling, several positions were investigated inside the previously modeled β-D-glucan structure, when testing the different orientations of the docking of AFB$_1$. There were minor differences between the energy values obtained at the different positions of both molecules during the setting up of the docking. AFB$_1$ seemed to be involved in a major stacking interaction due to van der Waals, which accounted for 93.4 % of the total energy of the docking. This chemical interaction between the AFB_1 core structure made up of an aromatic and a pyran cycle and the residues of β-D-glucopyranose ring (Figure 3), was responsible for a stabilising effect due to the geometrical symmetry and proximity between AFB1 and the cavities formed in the helical structure of β-D-glucans. DON was able to create chemical interactions with the modelled β-D-glucans despite a less stable stereochemistry compared with other mycotoxins because of its

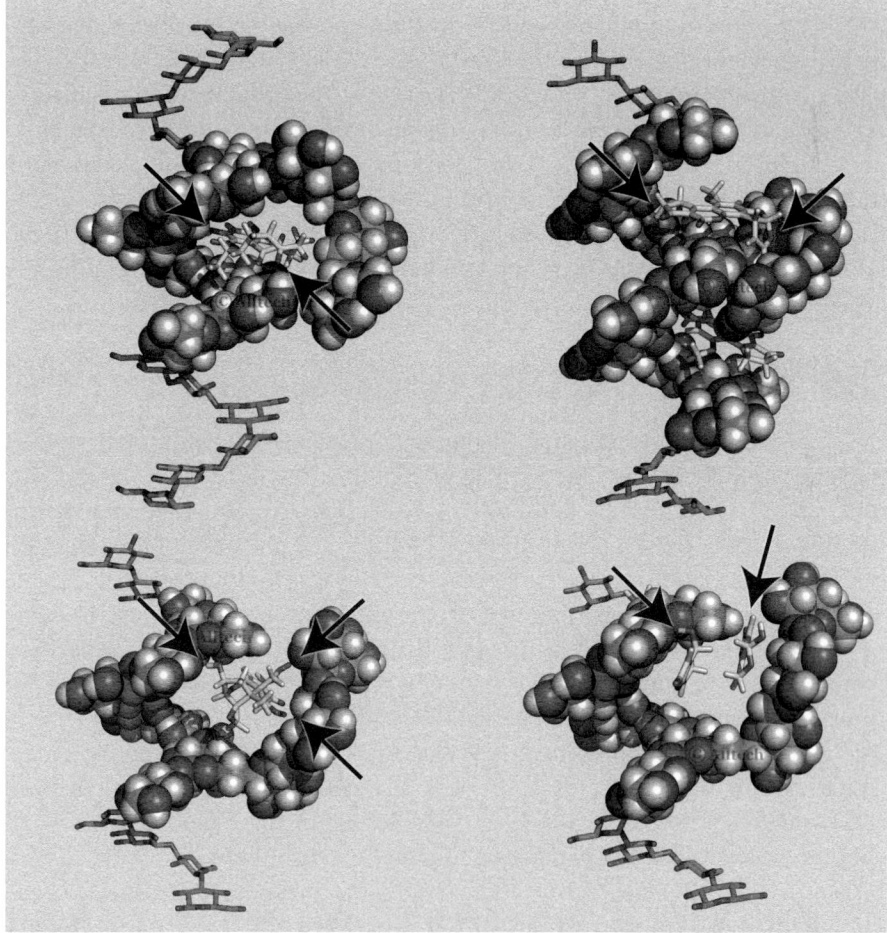

Figure 3. Computer-generated views of the docking of ZEA (top left), AFB1 (top right), DON (bottom left) and PAT (bottom right) into the single-helix of (1→3)-β-D-glucan branched with (1→6)-β-D-glucan side chains. Arrows indicate hydrogen bonds involved in the interactions.

tetracyclic scirpenol ring common to all trichothecenes (Figure 3). This weak stability of DON could be explained by the tetracyclic scirpenol ring and by the surrounding hydroxyl, epoxy and ketone groups. However, in spite of the lack of complementary geometry with β-D-glucans, the overall structure of DON was able to enter inside the binding site of the helix with or without the presence of the (1→6)-β-D-glucans side chain. The most stable conformation of the interaction could be clearly assigned to the formation of hydrogen bonds between the two hydroxyl groups of the DON molecule (Figure 1) and the hydroxyls of the β-glucopyranose residues. PAT had also an important stability in the β-D-glucan structure in different docking positions. It could be hypothesised that the low molecular size, the low steric bulk and the very planar conformation of its structure, gave to PAT the property of interacting in several manners and penetrating very deeply inside the single helical structure of β-D-glucan. However, only two positions were able to create electrostatic interactions leading to the formation of two hydrogen bonds and also giving the lowest values for the total energy of docking (Figure 3). These hydrogen bonds involved the hydroxyl group and the lactone group of the PAT molecule and the hydroxyl groups of the single-helix of the modelled β-D-glucans.

4. Conclusions

Basic science in biochemistry is required to complete and support in a meaningful way the advances in applied nutrition needed to extend our understanding of the beneficial role of *Saccharomyces cerevisiae* cell wall in limiting the toxicological impact of mycotoxins. The plasticity of the structure of β-D-glucan exhibiting diverse stereochemistry was undoubtedly responsible for the affinity of a large range of mycotoxins. Affinity rates varied widely between toxins due to their structural and physico-chemical disparities. Nevertheless, we concluded that β-D-glucans may have strong affinities for mycotoxins exhibiting 'aflatoxin-like', 'deoxynivalenol-like' or 'zearalenone-like' structures and represent a key tool for counteracting mycotoxin risk. The use of organic material from yeast cell walls such as Mycosorb® (Alltech Inc.) could therefore help sequester mycotoxins in the digestive tract when they are added to the diet and thus mitigate the harmful impact of mycotoxins on animals. Furthermore, considering their structural conformation, glucans are not implicated in interactions with nutrients. Finally, as organic materials, they are environmentally safe because of their biodegradable properties after excretion by the animal.

References

Ahokas, J., El-Nemazi, H., Kankaanpää, P., Mykkänen, H. and Salinen, S. (1998). A pilot clinical study examining the ability of a mixture of *Lactobacillus* and *Propionibacterium* to remove aflatoxin from the gastrointestinal tract of healthy Egyptian volunteers. *Review of Veterinary Medicine* **149:** 568.

Berthiller, F., Schuhmacher, R., Buttinger, G. and Krska, R. (2005). Rapid simultaneous determination of major type A- and B-trichothecenes as well as zearalenone in maize by high-performance liquid chromatography-tandem mass spectrometry. *Journal of Chromatography, A* **1062:** 209-216.

Catley, B.J. (1988). Isolation and analysis of cell walls. Chaper 8. In: Yeast; a practical approach (Ed. Campbell, J.H.D.). IRL Press, Oxford, Washington, USA, pp. 163-183.

Council for Agricultural Science and Technology (2003). *Mycotoxins: Risk in plant, animal and human systems.* Task Force Report 139. Ames, Iowa, USA.

Chuah, C.T., Sarko, A., Deslandes, Y. and Marchessault, R.H. (1983). Packing analysis of carbohydrates and polysaccharides. Part 14. Triple-helical crystalline structure of curdlan and paramylon hydrates. Macromolecules 16: 1375-1382.

Cordier, C., Gruselle, M., Jaouen, G., Hughes, D.W. and McGlinchey, M.J. (1990). Structures of zearalenone and zearalanone in solution: a high-field NMR and molecular modelling study. *Magnetic Resonance Chemistry* **28:** 835-845.

Corley, R.A., Swanson, S.P. and Buck, W.D. (1985). Glucuronide conjugates of T-2 toxin and metabolites in swine bile and urine. *Journal of Agriculture and Food Chemistry* **33:** 1085-1089.

Cotty, P.J., and Bhatnagar, D. 1994. Variability among atoxigenic *Aspergillus flavus* strains to prevent aflatoxin contamination and production of aflatoxin biosynthetic pathway enzymes, *Applied Environmental Microbiology* **60:** 2248-2251.

Dallies, N., Francois, J. and Paquet, V. (1998). Yeast functional analysis reports: a new method for quantitative determination of polysaccharides in the yeast cell wall. Application to the cell wall defective mutants of *S. cerevisiae. Yeast* **14:** 1297-1306.

Devegowda,G., Raju, M.V.L.N., Afzali, N. and Swamy, H.V.L.N. (1998). Mycotoxin picture world-wide: novel solutions for their counteraction. In: *Biotechnology in the Feed Industry, Proceedings of Alltechís 14th Annual Symposium* (Eds. Lyons, T.P. and Jacques, K.A.), Nottingham University Press, Nottingham, UK, pp. 241-256.

Diaz, D.E., Hagler, W.M., Hopkins, B.A. and Whitlow, L.W. (2002). Aflatoxin binders I: in vitro binding assay for aflatoxin B1 by several potential sequestering agents. *Mycopathologia* **156:** 223-226.

Engelhardt, G., Zill, G., Wohner, B. and Wallnofer, P.R. (1988). Transformation of the *Fusarium* mycotoxin zearalenone in maize cell suspension cultures. *Naturwissenschaften* **75:** 309-310.

Fleet, G.H. (1991). Cell walls. In: *The Yeasts*, Vol. 4 (Eds. Rose, A.H. and Harrison, J.S.). Academic Press, London, UK, pp. 199-271.

Grant, P.G. and Phillips, T.D. (1998). Isothermal adsorption of aflatoxin B1 on HSCAS clay. Journal of Agricultural and Food Chemistry 46: 599-605.

Jouany., J.-P. (2007). Methods for preventing, decontaminating and minimizing the toxicity of mycotoxins in feeds. Animal Feed Science and Technology 137: 342-362.

Kogan, G. (2000). (1→3)-β- and (1→6)-β-D-glucans of yeasts and fungi and their biological activity. In: *Studies in Natural Products Chemistry, Bioactive Natural Products (Part D)*, Vol. 23 (Ed. Rahman, A.). Elsevier Science B.V., The Netherlands, pp. 107-151.

Liu, Y., Walker, F., Hoegleninger, B. and Buchenauer, H. (2005). Solvolysis procedures for the determination of bound residues of the mycotoxin deoxynivalenol in Fusarium species infected grain of two winter wheat cultivars preinfected with barley yellow dwarf virus. Journal of Agricultural Food and Chemistry 53: 6864-6869.

Marchessault, R.H. and Deslandes, Y. (1979). Fine structure of (1→3)-β-D-glucans: curdlan and paramylon. Carbohydrate Research 75: 231-242.

Nakazato, M., Morozumi, S., Saito, K., Fujinuma, K., Nishima, T., and Kasai, N. (1990). Interconversion of aflatoxin B1 and aflatoxicol by several fungi. *Applied Environmental Microbiology* **56:** 1465-1470.

Nuessli, J., Putaux, J.-L., Le Bail, P., and Buleon, A. (2003). Crystal structure of amylose complexes with small ligands. International Journal of Biological Macromolecules 33: 227-234.

Panneerselvam, K., Rudino-Pinera, E. and Soriano-Garcia, M. (1996). 3,4,5,6,9,10-Hexahydro-14,16-dihydroxy-3-methyl-1H-2-benzoxacyclotetradecin-1,7(8H)-dione (Zearalenone). Acta Crystallographica 52: 3095-3097.

Pfohl-Leszkowicz, A. (2000). Risques mycotoxicologiques pour la santé des animaux et de l'homme. *Cahiers de Nutrition et de Diététique* **35:** 389-397.

Pittet, A. (1998). Natural occurrence of mycotoxins in foods and feeds – an update review. *Review of Veterinary Medicine* **149:** 479-492.

Popolo, L., Gilardelli, D., Bonfante, P. and Vai, M. (1997). Increase in chitin as an essential response to defects in assembly of cell wall polymers in the ggp1 delta mutant of S. *cerevisiae. Journal of Bacteriology* **179:** 463-469.

Richard., J., Thurston, J.R. and Pier, A.C. (1978). Effect of mycotoxins on immunity. In: *Toxins: animal, plants and microbial* (Ed. Rosenberg, P.). Persimmon, New York, USA, pp. 801-817.

Rotter, B.A., Thompson, B.K., Lessard, M., Trenholm, H.L., and Tryphonas, H. (1994). Influence of low-level exposure to *Fusarium* mycotoxins on selected immunological and haematological parameters in young swine. *Fund Applied Toxicology* **23:** 117-124.

Schneweis, I., Meyer, K., Engelhardt, G. and Bauer, J. (2002). Occurrence of zearalenone-4-β-D-glucopyranoside in wheat. Journal of Agriculture and Food Chemistry 50: 1736-1738.

Schoental, R. (1984). Mycotoxins and the Bible. *Perspectives in Biology and Medicine* **28:** 117-120.

Schoental, R. (1991). Mycotoxins, porphyrias and the decline of the Etruscans. *Journal of Applied Toxicology* **11:** 453-454.

Schoental, R. (1994). Mycotoxins in food and the plague in Athens. *Journal of Nutrition and Medecine* **4:** 83-85.

Scott, P.M. (1998). Industrial and farm detoxification processes for mycotoxins. *Review of Veterinary Medicine* **149:** 543-548.

Smith, T.K., McMillan, E.G. and Castillo, J.B. (1997). Effect of feeding blends of *Fusarium* mycotoxin-contaminated grains containing deoxynivalenol and fusaric acid on growth and feed consumption of immature swine. *Journal of Animal Science* **75:** 2184.

Stoev, S.S, Goundasheva, D., Mirtcheva, T. and Mantle, P.G. (2000). Susceptibility to secondary bacterial infections in growing pigs as an early response in ochratoxicosis. *Experimental and Toxicological Pathology* **52:** 287-296.

Trenholm, H.L., Hamilton, R.M.G., Friend, D.W., Thompson, B.K. and Hartin, K.E. (1984). Feeding trials with vomitoxin (deoxynivalenol)-contaminated wheat: Effect on swine, poultry, and dairy cattle. *Journal of the American Veterinary Medical Association* **185:** 527-531.

Whitlow, L.W., Diaz, D.E., Hopkins, B.A. and Hagler, W.M. (2000). Mycotoxins and milk safety: the potential to block transfer to milk. In: *Proceedings of Alltech's 16th Annual Symposium* (Eds. Lyons, T.P. and Jacques, K.A.), Nottingham University Press, Nottingham, UK, pp. 391-408.

Yiannikouris, A., Poughon, L., Cameleyre, X., Dussap, C.-G., François, J., Bertin, G., and Jouany, J.-P. (2003). A novel technique to evaluate interactions between *S. cerevisiae* cell wall and mycotoxins: application to zearalenone. Biotechnology Letters 25: 783-789.

Yiannikouris, A., François, J., Poughon, L., C-G Dussap, Bertin, G., Jeminet, G. and Jouany, J.-P. (2004a). Adsorption of zearalenone by β-D-glucans in the *S. cerevisiae* cell wall. Journal of Food Protection 67: 1195-1200.

Yiannikouris, A., François, J., Poughon, L., Dussap, C.-G., Bertin, G., Jeminet, G. and Jouany, J.-P. (2004b). Alkali-extraction of β-D-glucans from *S. cerevisiae* cell wall and study of their adsorptive properties toward zearalenone. Journal of Agriculture and Food Chemistry 52: 3666-3673.

Yiannikouris, A., François, J., L Poughon, L., Dussap, C.-G., Jeminet, G., Bertin, G., and Jouany, J.-P. (2004c). Influence of pH on model β-D-glucans complexation toward zearalenone. Journal of Food Protection 67: 2741-2746.

Yiannikouris, A., André, G., Buléon, A., Jeminet, G., Canet, I., François, J., Bertin, G., and Jouany, J.-P. (2004d). Comprehensive conformational study of key interactions involved in zearalenone complexation with β-D-glucans. Biomacromolecules 5: 2176-2185.

Yiannikouris, A., André, G., L Poughon, François, J., Dussap, C.-G., Jeminet, G., Bertin, G., and Jouany, J.-P. (2006). Chemical and conformational study of the interactions involved in mycotoxins complexation with ß-D-glucans. Biomacromolecules 7: 1147-1155.

Zhou, B. Li, Y., Gillespie, J., He, G.-P., Horsley, R. and Schwarz, P. (2007). Doehlert matrix design for optimization of the determination of bound deoxynivalenol in barley grain with trifluoroacetic acid (TFA). *Journal of Agriculture and Food Chemistry* **55:** 10141-10149.

Salmonella in modern pig production: here today, gone tomorrow?

Thomas Blaha
University of Veterinary Medicine Hannover, Foundation, Field Station for Epidemiology, 49456 Bakum, Germany

1. Introduction

Salmonella is an important cause of food-borne illness in humans. Farm animals and foods of animal origin form an important source of human *Salmonella* infections. To address this fact, which is unanimously accepted in the scientific community throughout Europe, the EU Commission issued two major regulations: Directive 2003/99/EC demanding a regular and EU-standardised monitoring and reporting, and Regulation (EC) 2061/2003 demanding the systematic reduction of the occurrence of zoonotic pathogens in food animals, and thus, in the food chain. In the following section, the current knowledge on *Salmonella* in pig production is explained.

2. *Salmonella* in pigs: here today?

Yes, there is no doubt: *Salmonella* is 'here today' in modern pig production. This means that the question mark is not to question whether *Salmonella* is here today (we know it is), but only to question the 'today', or in other words: the question is whether this is a new threat to human health? There is ample evidence that *Salmonella* in food animals (and thus in food of animal origin) is not a public health concern of only today, but it has obviously been around for many decades prior to today. If *Salmonella* has been around for so long, why do we focus today so much more intensely than 'yesterday' on surveying and controlling *Salmonella* in pig production? There are two explanations:
a. Until recently, the so called 'classical' food safety risks carried by meat (tuberculosis, brucellosis, tape worms, trichinellosis, etc.) were prevailing and the major targets of any control programme (meat inspection being the major tool) – having these threads more or less under control, *Salmonella* can now be targeted.
b. There has been and still is under way a remarkable consolidation in food production: herds are getting larger and especially slaughter and meat processing takes place nowadays in huge production facilities – this does not

mean necessarily more salmonella contaminations occurs (introductions of *Salmonella* into the food chain at any level), but that they lead undoubtedly to more severe consequences, i.e. more outbreaks with more people affected per outbreak. It can even be assumed that the frequency of human cases were always in the range of today's prevalence, but most of the sporadic cases of the past were less visible and less reported than the sometimes spectacular outbreaks of today.

In general, scientists and epidemiologists agree on the following assessments: (1) up to 80% of the *Salmonella* contamination of food results from *Salmonella*-positive food animals (either they are the direct source or at least the reservoir, (2) on average, out of the human salmonellosis cases due to food of animal origin, about 60% are due to poultry products, 30% are due to pork and 10% are due to bovine products. Although these percentage estimates apply to all European countries, there are significant differences in the national prevalence of *Salmonella* infections in food animals: Whereas Sweden, Finland and Norway can be proud of a very low *Salmonella* prevalence in all food animal species, i.e. also in pigs, the prevalence of *Salmonella* infections in the national food animal populations of the other EU member states is significantly higher (at least 10 times higher than in these three Scandinavian countries). Only Denmark has a remarkably lower *Salmonella* prevalence than the so called 'higher prevalence countries' (about half of their prevalence). The resultant *Salmonella* prevalence classification of the EU member states into 'low prevalence', 'medium prevalence' and 'high prevalence' countries has been confirmed exactly by the baseline studies for laying hens and slaughter pigs (according to 2003/99/EG).

As for the baseline survey in slaughter pigs (EFSA, 2008), the sampling of slaughter pigs took place between October 2006 and September 2007. All participating Member States and Norway sampled ileocæcal lymph nodes from the selected slaughtered pigs. In total 18,751 slaughter pigs were sampled and 18,663 valid lymph node samples were collected.

Twenty-four of the 25 participating Member States isolated *Salmonella* spp., which resulted in an EU observed prevalence of *Salmonella*-positive slaughter pigs of 10.3%. This means that in the European Union at the point of slaughter around one in ten slaughter pigs were estimated to be infected with *Salmonella* in the lymph nodes. The *Salmonella* prevalence in these slaughter pigs varied widely amongst the Member States, from 0.0% to 29.6%. All 24 *Salmonella*-positive Member States isolated *Salmonella Typhimurium* and 20 detected *Salmonella Derby,* which are two very common serovars found in *Salmonella* infection cases in humans. This resulted in an estimated EU observed prevalence of 4.7% for *Salmonella Typhimurium*, varying from 0.0% to 16.1% within the Member States, and of 2.1% for *Salmonella Derby*, varying from 0.0% to 6.5%. Figure 1 shows the prevalence of carcasses with ileocaecal lymph nodes positive for *Salmonella* spp. per country.

The diversity of isolated *Salmonella* serovars in slaughter pig lymph nodes was large, and in total, 94 different serovars were isolated in the European Union. The five most frequently isolated *Salmonella* serovars from lymph nodes in the European Union were respectively in decreasing order *S. Typhimurium*, *S. Derby*, *S. Rissen*, *S. Enteritidis* and *S.* 4,[5],12:i:-. All these serovars, with the exception of *S. rissen*, are frequent causes of *Salmonella* infections in humans within the European Union. *Salmonella Typhimurium* and *Salmonella Derby* serovars were highly predominant in lymph nodes; *S. Typhimurium* was the most

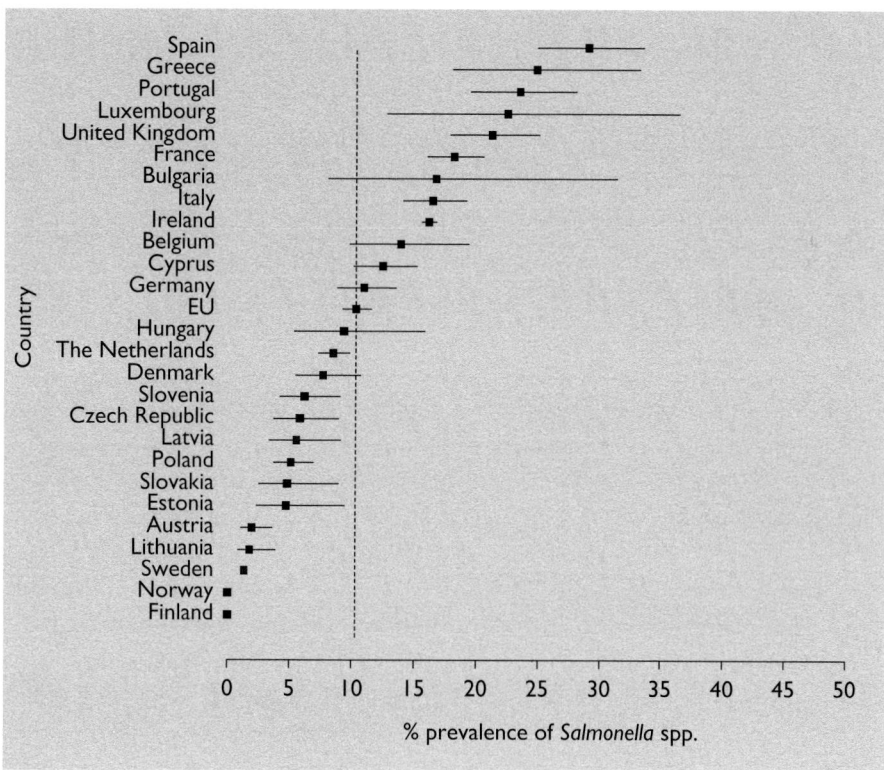

Figure 1. Observed prevalence of slaughter pigs infected with Salmonella spp. in lymph nodes, with 95% confidence intervals, in the EU and Norway, 2006-2007.

common serovar, detected in 40.0% of the *Salmonella*-positive slaughter pigs and reported by all 24 *Salmonella*-positive Member States. *S. Derby* accounted also for an important proportion of positive lymph nodes (14.6%) and was reported by 20 *Salmonella*-positive Member States. Figures 2, 3, and 4 show the prevalence of carcasses with ileocaecal lymph nodes positive for *Salmonella Typhimurium* (Figure 2), *Salmonella Derby* (Figure 3) and *Salmonella* serovars other than *S. Typhimurium* and *S.Derby* (Figure 4) per country.

2.1. General summary of the results

It is important to note that the absence of any *Salmonella* from the tested samples does not imply that a Member State is *Salmonella* free, as firstly the detection method has a sensitivity of less than 100%, so false negative results

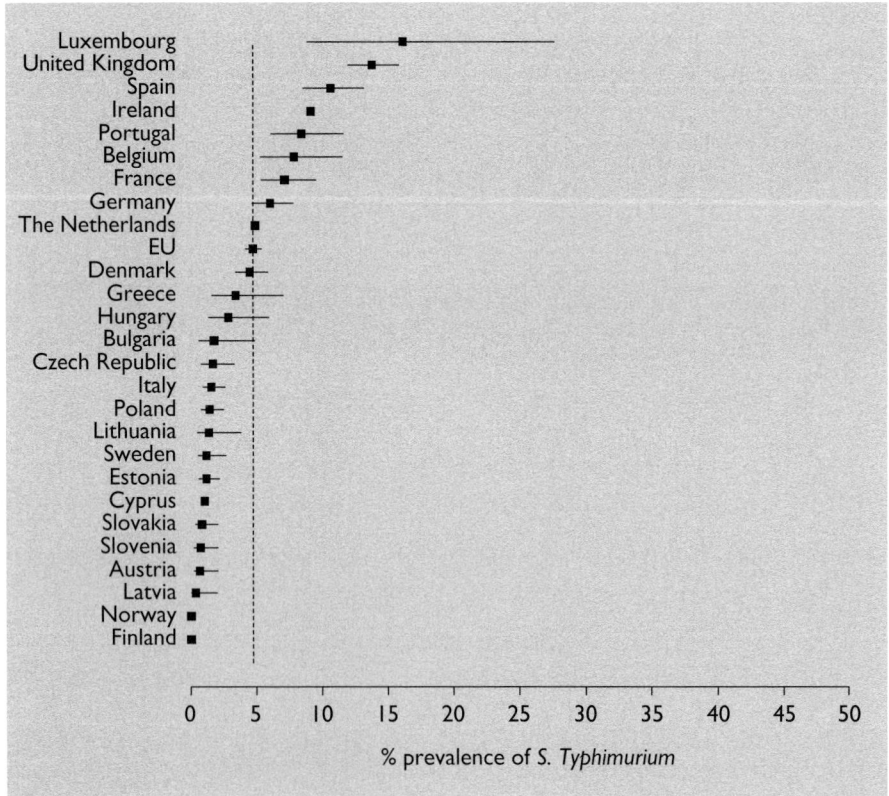

Figure 2. Observed prevalence of slaughter pigs infected with S. typhimurium in lymph nodes, with 95% confidence intervals, in the EU and Norway, 2006-2007 (EFSA, 2008).

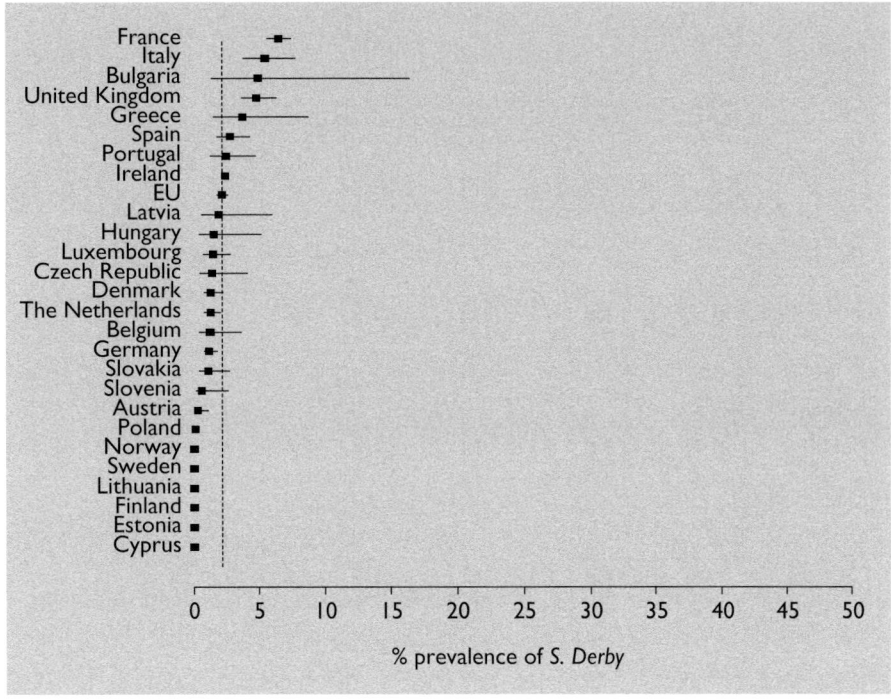

Figure 3. Observed prevalence of slaughter pigs infected with S. derby *in lymph nodes, with 95% confidence intervals, in the EU and Norway, 2006-2007 (EFSA, 2008).*

are plausible. Secondly, the prevalence within the Member States may be too low for even one positive animal to be detected with the sample size that was used. The EU wide, weighted prevalence of *Salmonella* infection in lymph nodes was 10.3%, with a 95% confidence interval of 9.2%-11.5%. Within the EU Member States, the prevalence varied between 0.0% and 29.6%. This can be interpreted as showing that approximately one in ten pigs slaughtered in the EU was infected with *Salmonella* when slaughtered. This infection may have arisen on the farm of origin or at any time during transport to slaughter or lairage. The lymph nodes results showed that about half of the EU Member States had a *Salmonella* prevalence in lymph nodes above the EU average, while the other half had prevalence below the EU mean. This was roughly speaking also the case for *S. Typhimurium*, but less true for *S. Derby* and for serovars other than these latter two for which fewer Member States had figures above the EU mean. Finland and Norway did not report any isolate from lymph nodes.

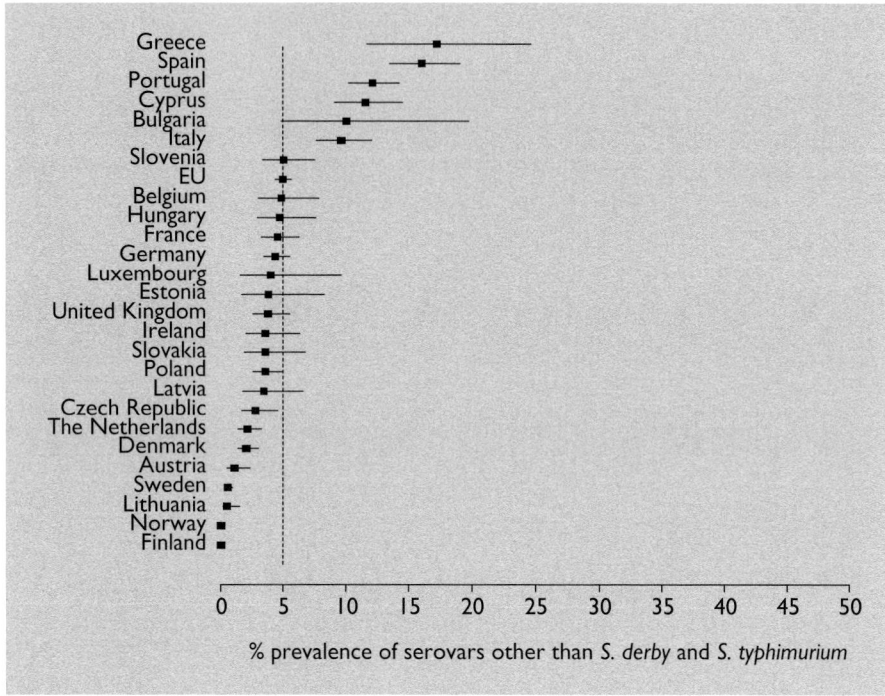

Figure 4. Observed prevalence of slaughter pigs infected with serovars other than S. typhimurium or S. derby in lymph nodes, with 95% confidence intervals, in the EU and Norway, 2006-2007 (EFSA, 2008).

It is noteworthy that although there was a large variation in the slaughter pig *Salmonella* prevalence, the serovar distribution was not remarkably varying between the EU Member States, because two specific *Salmonella* serovars, *S. typhimurium* and *S. Derby*, which are two very common serovars found in *Salmonella* infection cases in humans, accounted for a major part of the positive findings at the EU-level and for most *Salmonella*-positive EU Member States. Indeed, all 24 *Salmonella*-positive EU Member States isolated *Salmonella typhimurium* (EU observed prevalence of 4.7%) and 20 detected *Salmonella derby* (EU observed prevalence of 2.1%), which are two very common serovars found in *Salmonella* infection cases in humans. Both of these serovars have also been isolated from other species of domestic livestock, including poultry and have also been implicated in cases of clinical salmonellosis in humans.

Finally, the remarkably low *Salmonella* prevalence in Finland, Sweden and Norway, as well as the relatively low Salmonella prevalence in slaughter pigs in

Denmark compared to those EU Member States that have a quite intensive pig production needs some explanation:

1. Sweden, Finland and Norway (the so called 'Low Prevalence Countries'): The reason for the low prevalence in these three countries has nothing to do with the northern climate (as some may assume), but exclusively with the very strict Salmonella monitoring and control programmes that all there countries have implemented since the 60's (!) of the last century. The reason for that was a very severe outbreak of human salmonellosis due to *Salmonella typhimurium* in the late 50's in Sweden. The outbreak strain was obviously highly pathogenic, since there were many fatal cases throughout Sweden and hundreds of people had to be hospitalised. This outbreak was kind of a 'national wake up call' and led to the design of a strict *Salmonella* monitoring and control programme covering feed, animal production, and slaughter and food processing facilities. All monitoring has been done and still is done by means of bacteriological testing. The result is that today the three 'low prevalence countries' can claim that their human salmonellosis prevalence is significantly lower than that of all other EU Member States, and even more, that about 90% of the still occurring salmonellosis cases are 'imported' cases, i.e. cases that are due to an infection that has taken place abroad and became apparent after the return of the infected persons to their home country.

2. Denmark (the so called 'Medium Prevalence Country'): In 1993, Denmark experienced a major outbreak of human salmonellosis due to *Salmonella infantis*. The source of this outbreak was immediately and doubtlessly traced to pork (the outbreak was caused by a strike in the slaughter plant leading to an overflow of carcasses in the coolers and to drastic breakdowns of the slaughter plant hygiene. This incident which resulted in more than 1,000 people hospitalised in Copenhagen, which threatened Denmark's pork export, led to a joint effort of the state of Denmark and the pork industry that resulted in the Danish national Salmonella Monitoring and Control Programme. The core of this programme is the monitoring programme based on serological testing of random samples (meat juice) per pig herd, which was used to categorise the Danish pig herds according to the estimated risk of carrying *Salmonella* into the slaughter plants (the more pigs in the herd sample turned out to be carriers of *Salmonella* antibodies, the higher the risk that live animals of the herd in question carry *Salmonella* in their guts, tonsils and lymph nodes). The programme led in the course of about 10 years to a significant reduction of human salmonellosis that can be traced to pork.

In all other EU Member States, the prevalence of *Salmonella* in pigs is more or less comparably higher than that of the Scandinavian countries and Denmark. This is also true for those countries that have already started national or regional *Salmonella* monitoring programmes (e.g. Germany, UK, Ireland, The Netherlands) before the EU 'zoonosis regulations' had been issued. The reason that there has been so far not yet a statistically proven reduction of the *Salmonella* prevalence in these countries lies in the fact that obviously several years are needed before any monitoring and controlling of *Salmonella* in food animals leads to a measurable decrease of human salmonellosis cases.

In the following sections, before answering the second question (*Salmonella* in pigs – gone tomorrow?), *Salmonella* as bacterial species, and the risk factors for *Salmonella* in the pork chain, especially in the pre-harvest phase (feed production and feeding, animal husbandry, transport and lairage) are explained.

3. *Salmonella*: the nomenclature

The bacterial genus *Salmonella* has been named after Dr. D.E. Salmon in recognition of his work on animal diseases in the USA and southern America (Salmon and Smith, 1886). The genus *Salmonella* contains two species: *Salmonella enterica* and *Salmonella bongori* (Reeves *et al.*, 1989)). The species *Salmonella enterica* contains six subspecies of which the subspecies *enterica* contains many serovars that are important in pigs (Grimont *et al.*, 2000). Specific serovars of *Salmonella* (*S.*) are given names such as *Salmonella enterica* subsp. *enterica* serovar *dublin*, *Salmonella (S.) dublin* for short, usually named after the location where they were isolated for the first time. Serovars that are important in pigs are for instance *Salmonella (S.) Thyphimurium, S. Choleraesuis, S. panama, S. infantis, S. derby, S. hadar, S. london*, and so on (Devriese, 1995; Baggesen *et al.*, 1996; Geue *et al.*, 1997); Van der Wolf and Peperkamp, 2001).

Salmonellae can be divided in host-adapted serovars and those that have a much broader host range. *Salmonella choleraesuis* can be considered as host adapted to pigs. This is due to the fact that *S. choleraesuis* is a primary pathogen for pigs and only seldom infects other species including man. However, if *S. choleraesuis* infects humans, it can result in a severe disease, even in a fatal infection given the right circumstances. On the other hand, non-host adapted *Salmonellae* can infect a wide range of species including man who potentially falls ill – the most familiar example is *Salmonella thyphimurium* (Van Aartsen, 1997). These non-host specific *Salmonellae* can occasionally cause clinical disease in pigs (as a rule in cases there is already another enteric disorder or infection), but mostly pigs are healthy carriers of these *Salmonella* spp. After infecting the host, *Salmonella* spp. are excreted in large numbers in the faeces which results in a contamination of the environment resulting in the spread to other hosts (Murray, 1991).

4. *Salmonella* in the pork production chain

In industrialised countries such as in the EU member states the scheme according to which pork is produced is often represented as a pyramid. At the top pure breed pig herds are maintained to generate the genetic input for multiplying and finishing pigs. As a result of the immense capacity of pigs to reproduce, only a limited number of breeding animals are needed to produce millions of finishing pigs per year, which form the bottom of the pyramid (in December 2004, there were about 151 million pigs in the EU). The different stages of the production chains within this pyramid (breeding, rearing weaner pigs and grow-finish) is mostly performed in a multitude of different herds resulting in a huge transportation network of live pigs all across Europe. At any stage of production *Salmonella* can come into these herds and infect pigs. Sources of infection are for example (not in order of importance) pigs that are introduced into the herd, feed, the farm environment (including wild animals and/or their feaces), transport vehicles, visitors, rodents, rats, flies, pets, and birds. Once introduced into the herd *Salmonella* can remain there for many years infecting generation after generation of pigs (Van der Wolf *et al.*, 2001a). Within herds, two types of transmission are distinguished; horizontal and vertical. Horizontal transmission is probably most common and involves the direct or indirect faecal-oral spread from one animal to another with a wide range of methods of mechanical transport in between. Farmer's hands and boots, tools, pest animals and manure lorries are only a few of the examples. In Denmark for example, it was found that farms that share farm machinery are at a higher risk of getting contaminated with a certain type of *Salmonella* than farms that don't. The other type of transmission is vertical transmission. This transmission is best characterised passing *Salmonellae* from the sow to her piglets, which then transfer the *Salmonella* to the weaner barn and further to the grow-finish unit. In contrast to poultry were *Salmonella* Enteritidis can infect the egg from which an infected chick is hatched, piglets only get infected after birth by taking up faecal material contaminated with *Salmonella*. Antibodies in the colostrum of the sow have the potential to protect the piglets against infection. This is the biological basis for the fact that vaccinating the sow to protect her offspring can be part of possible intervention measures (see section 5.4).

After the primary production phase (from feed production through rearing the animals up to transporting them to slaughter) follows the phase in which finishing pigs and culled animals from other production stages (e.g. sows no longer used for breeding) are transported to the slaughterhouse, kept in holding areas, lairage, and after that stunned and killed. As mentioned above transport lorries can contaminate batches of pigs one after the other (Hurd *et al.*, 2002). The same is even more so for the lairage (Swanenburg *et al.*, 2000). Many batches of pigs are held here that have a higher risk of shedding *Salmonella* as a result of fasting at the farm, stress of driving, mingling pigs with pigs from other herds and stress during unloading. During holding, the pigs are showered by sprinklers to calm them down and clean them. As a result slurry is formed on the floor of the lairage which is contaminated with *Salmonella* in most instances. After oral or nasal uptake of *Salmonella,* dissemination through the entire body occurs within two to three hours resulting in shedding of *Salmonella* which were acquired during mingling, transport or holding (Hurd *et al.*, 2001). As a result potentially many more pigs are contaminated with *Salmonella* at the moment of slaughter on their skin and in their nasopharynx and in their entrails than primarily infected pigs from the farm (Letellier *et al.*, 1999; Davies *et al.*, 1999; Beloeil *et al.*, 2004).

5. Risk factors and interventions during the pre-harvest phase

Risk factors for *Salmonella* infection in pigs are related to (a) the introduction of *Salmonella* into the herd, and (b) the spread of *Salmonellae* within the herd, as well as to the capability of pigs to withstand infection and to minimise the multiplication and shedding of the pathogen.

5.1. Risk factors for introduction of Salmonella into the herd

- Bringing infected pigs into the herd is probably the most important source of introduction of *Salmonella.*
- Feed, or specific feed components if feed components are bought separately, can be a source of *Salmonella.*
- All visitors to a herd can bring *Salmonella* with them.
- Rats, mice, flies, birds, foxes, raccoons, cats, dogs, etc., but also other farm animals can get contaminated and infected with *Salmonella* and spread this *Salmonella* to other animals, e.g. pigs.
- Husbandry systems where pigs are kept outside total confinement (pasture, free range, etc.) are at an increased risk of getting infected with *Salmonella.*

- Sharing farm equipment can introduce *Salmonella* into the herd.
- Water, other than from the municipal supply (e.g. wells on the farm), can introduce *Salmonella*.

5.2. Risk factors for the spread of Salmonella *within the herd*

- Improper cleaning and disinfection procedures promote the spread of *Salmonella*.
- Continuous flow or deficiencies in the all-in/all-out management increase the spread of *Salmonella*.
- Lack of age group separation supports the transmission of *Salmonella* from older to younger pigs and vice versa.
- Lack of solid pen separation walls also supports the transmission of *Salmonella* from one pen to another.
- Pest animals can spread *Salmonella* within the herd.

5.3. Risk factors that increase the susceptibility of pigs for Salmonella *infection*

- Infection with *Ascaris suum*.
- Infection with other enteric pathogens, e.g. *Brachyspira hyodysenteriae* and *Lawsonia intracellularis*.
- Using antibiotics and antimicrobial growth promoters that are disturbing the microbial balance in the gut.

5.4. Intervention methods for preventing and minimising the introduction of Salmonella *into the herd and the spread of* Salmonella *within the herd*

Bringing infected pigs into the herd is probably the most important source of introduction of *Salmonella*. The problem with preventing this risk is that mostly infected animals are clinically healthy who start shedding *Salmonella* after transport in the new herd. Buying certified *Salmonella*-free pigs is only possible when buying specified pathogen-free (SPF) animals where *Salmonella* is part of the definition of SPF, which is exceptional and not feasible for growers and finishers. Exception to this is the situation in Sweden and Finland where there is virtually no *Salmonella* present. However, it is very sensible to try to intervene at the earliest stage, e.g. in weaners, because these are smaller animals that consume less feed and water (see later). From research in the U.K. it became clear that sow herds that do not buy replacement gilts have a lower chance of being infected with *Salmonella* then herds that do buy replacement gilts.

Decontaminating potentially contaminated feed or high risk feed components, e.g. with formic acid or by pelleting will reduce contamination considerably (Edel *et al.*, 1974; Fedorka-Cray *et al.*, 1997; Harris *et al.*, 1997).

All visitors to a herd can bring *Salmonella* with them. For this reason (but more importantly for OIE list A diseases as well as for any other disease that is to be prevented from being introduced into the herd) visitors should have to change their footwear and wear coveralls and wash their hands (ideally taking a full shower) before entering a herd (Van der Wolf *et al.*, 2001c). It is important that the care taker of the pigs obeys by these rules as well as he/she is the most frequent visitor to the herd. Equally important are tools or machinery that are (temporarily) brought into the herd. All tools and machinery should be thoroughly cleaned and disinfected before they are brought into the herd.

Rats, mice, flies, birds, foxes, raccoons, cats, dogs, etc. but also other farm animals can get contaminated and infected with *Salmonella* and spread this *Salmonella* to other animals, e.g. pigs. Biosecurity of buildings should include measures which prevent these animals coming into the herd and well established control schemes should deal with pests that are not kept out. Pigs should be kept separate from other farm animals. Husbandry systems where pigs are kept outside total confinement (pasture, free range, etc.) are at an increased risk of getting infected with *Salmonella* (Van der Wolf *et al.*, 2001d). Control under these circumstances will be very difficult as a result of the continuous exposure. Any population prevalence estimate should be stratified to monitor these kinds of herds separately from total confined housing.

An effective way to prevent the spread of *Salmonella* is using all-in/all-out with thorough cleaning and disinfection between batches. This is demonstrated by the difference in prevalence between sows, which are kept in a sort of continuous system and finishers which are kept under all-in/all-out conditions (Van der Wolf *et al.*, 2001d). Moreover it is possible to rear growers and finishers free from *Salmonella* that come from sow herds that are infected with *Salmonella* (Dahl *et al.*, 1997) using all-in/all-out and thorough cleaning and disinfection. *Salmonella* infections that are horizontally transmitted within (finishing) herds as a result of not implementing complete all-in/all-out and thorough cleaning and disinfection between batches is probably the most important source of contamination and infection in pigs. As cleaning and disinfecting is hard and dirty work it is not very popular among pig farmers. However, with proper coaching an acceptable result is attainable.

Closed pen separation can prevent transmission of faecal material from one pen to another and therefore reduce the speed with which *Salmonella* can spread through a herd. Proper pest control strategies will reduce the pest burden within a herd and will result in a lower transmission rate by pest animals. Tools such as shovels and brooms, but also boots should preferably be allocated to a particular barn or age group of pigs. Passages from one age group to another or from one barn to another should preferably be equipped with hygiene facilities.

Antimicrobial treatments or antimicrobial growth promoters can disrupt the gut flora and increase the susceptibility of pigs for *Salmonella* infection (lit ref tylosine). Unnecessary or prolonged antimicrobial treatment should be prevented. Infection with *Ascaris suum* can increase the susceptibility of pigs for *Salmonella* infection. Effective deworming schemes should be applied. Other enteric diseases that damage the gut wall, e.g. infection with *Brachyspira hyodysenteriae*, increase the susceptibility of pigs for *Salmonella*-infection. Veterinary health management should be aimed at reducing enteric diseases in pigs.

Feed is not only a potential source of *Salmonella* contamination, but also a way by which *Salmonella* intervention can be reached. There are several possibilities:
- As shown by Edel *et al.* (1970) pelleting can strongly reduce *Salmonella* contamination of complete feeds.
- In Denmark a wide range of research is dedicated to the fact that meal feeding is linked to a lower *Salmonella* infection status than feeding pelleted feed. A prerequisite for this to function in light of the first point mentioned above is that the ingredients of meal feed are *Salmonella* free to begin with. Given this condition, meal creates a less watery substance in the stomach of the pig in comparison to pellet feeding, which in turn results in a lower pH as a result of more gastric acid production and longer retention time in the stomach as a result of which *Salmonella* are killed in the stomach of the pigs fed with meal feed. In short, one can say that meal feed enhances the gastric barrier for *Salmonella* in pigs. Feeding meal is recommended in Denmark as a method to control *Salmonella* in pigs.
- Feeding pigs with fermented liquid feed (FLF) containing by-products from the human food industry strongly protects against *Salmonella* in pigs. Herds fed this kind of feed can stay *Salmonella* free for years (Van der Wolf *et al.*, 2001a). A drawback of such a feeding system is that it requires a feed mixing kitchen with computer operated pumps and pipelines which requires quite a large investment. Only in finishing herds with more than 1200 to 1500

places would such a system be financially viable. Availability of such by-products, which can be supplied with HACCP certificates, is a prerequisite. However, when feasible this system has four major advantages:

1. The system recycles products which would otherwise have to be dumped at considerable costs and would burden the environment.
2. Cost saving for the pig farmer because such feeds are usually cheaper than compound feeds.
3. Improved pig health and performance resulting in higher financial gain for the farmer.
4. Improved *Salmonella*-status of these herd in comparison to herds that feed compound feed resulting in an improvement of public health.

- It is also possible to add organic acids (e.g. propionic acid) to the feed and/or drinking water or potassium diformate (Kamphues, 2006) to the feed of pigs. Although the effect is not as strong as with FLF, *Salmonella* can be reduced with these additions (Van der Wolf *et al.*, 2001b).
- Increasing the percentage of raw fibre and barley in roughly ground compound feed increases the enzymatic digestion in the large bowel, which reduces the multiplication in the distal parts of the gut (Kamphues, 2006).
- An oral attenuated live vaccine based on *Salmonella thyphimurium* is available in Germany. Experiences with this vaccine are limited and the vaccine is not available in all European countries at this moment. Vaccination at an early stage of life (after weaning) would not interfere with serological detection of antibodies against *Salmonella* at the end of the finishing period. A special serological test has been developed to distinguish between vaccinated and naturally-infected animals. Disadvantage of such vaccines is that they are serovar specific and offer probably only limited cross protection to infection with *Salmonella*'s from the same serogroup and no protection against infection with *Salmonella*'s from other serogroups. However, vaccination could very well play an important role in the intervention of *Salmonella* in high prevalence herds (Lumsden and Wilkie, 1992; Ortmann, 1999; Springer, 2001).

6. Re-evaluating some of the so far undoubted beliefs on *Salmonella* in pig herds

6.1. The importance of residual Salmonella spp. at farm level

Due to a recent study at the Field Station for Epidemiology of the University of Veterinary Medicine Hannover (Bode, 2007; Bode *et al.*, 2008), we are able to say that so far we have overestimated the permanent introduction of *Salmonella* into farms that are identified as high risk farms, and that we have underestimated the residual Salmonella reservoirs, that even exist after seemingly intensive cleaning and disinfection procedures in an all-in/all-out setting.

In this study the in-depth analysis of the reasons for a very high *Salmonella* load of an otherwise high-health and well-managed pork production system and the measures that were taken to reduce the prevalence of *Salmonella* are described. The investigations took place in a closed swine production system consisting of a breeding herd (680 sows), a nursery and three connected finishing herds which have shown a high *Salmonella* seroprevalence for a long time (finishing herd 1: 52.9%, finishing herd 2: 73.8%, finishing herd 3: 91.2%). In the past some of the gilts were *Salmonella* seropositive and *S. Derby* was found in piglet diarrhoea.

Based on the historical data, the assumption was the continuous introduction of *Salmonella* into the three finishing herds via infected weaners and grow-finishers.

The first phase of the study was to identify the *Salmonella* infection sources. From 42 sows blood was drawn in late gestation, three piglets per sow were earmarked and were followed until slaughter. At various points in time blood samples of the earmarked pigs were collected. All together, 751 serological samples were investigated for *Salmonella* antibodies. Simultaneously, 538 diverse bacteriological samples such as faeces, environmental samples and slaughter samples (tonsils, *Ln. mandibularis*, *Lnn. iliaci*) were cultivated for *Salmonella*. Based on these results specific measures for reducing *Salmonella* were implemented in the nursery and the three finishing herds. In the second phase of the study, the effect of the measures taken to reduce the *Salmonella* load was measured. Sixty randomly chosen and earmarked weaners were divided into the same three finishing barns from phase 1 and blood samples were collected at five points in time. All together, 368 serological samples and 556 bacteriological samples were investigated for *Salmonella*.

In phase 1 the breeding herd was not *Salmonella* free, but showed only a low *Salmonella* load (faeces: 5.9%, environmental: 0%). In the nursery the *Salmonella* load was similarly low (faeces: 2.9%, environmental: 0%). However, high bacteriological *Salmonella* rates in the finishing herds 1, 2 and 3 were seen in faeces (56.3%; 25.0%; 15.6%) and in environmental samples (31.1%; 7.5%; 5.0%) (Figure 5), and a steep increase of the sero-prevalence in the finishing herds was observed (Figure 6).

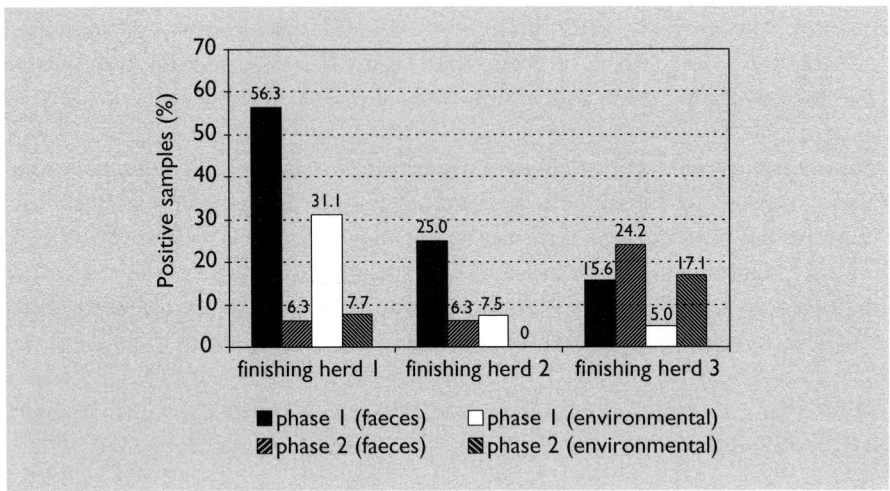

Figure 5. The bacteriology of the finishing herds 1-3.

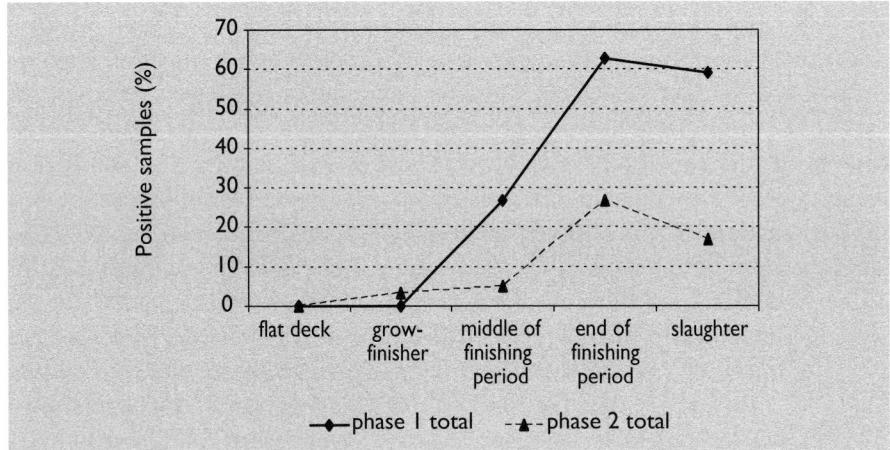

Figure 6. The serology of the finishing herds 1-3.

According to these results, the following measures for reducing *Salmonella* were implemented: (1) Intensive cleaning and disinfection of areas accessible to pigs; (2) Additional disinfection of areas with rare pig contact; (3) Cleaning and disinfection of areas that are not regularly included in cleaning and disinfection; (4) Improving the anterooms with strict separation between normal and farm clothes and boots; (5) Installing boots and disinfection mats inside each pig barn; (6) Disinfection of well water with chlorine dioxide; (7)

Switch to municipal water supply instead of well water; (8) Rough grinding of grain compo-nents; (9) Increase the amount of barley in the ration and (10) Addition of 0.6-1.2% potassium diformate to the feed.

After having implemented these measures, a remarkable reduction of *Salmonella* detected by culture could be seen in the finishing herd 1 and 2, but an increase in finishing herd 3 (Figure 5). The percentage of *Salmonella* antibody positive samples of all three herds was significantly reduced during the finishing period until slaughter (Figure 6).

In this production chain clearly the residual contamination of the finishing barn equipment (e.g. drinkers, feeding troughs) and indirect vectors (e.g. boots, tools, transport vehicles, loading ramps, fans) were (in contrast to the assumption) the major reason for the high *Salmonella* load in the three finisher herds. It could be proven that the traditional cleaning and disinfection procedures are not efficient enough to get rid of residual *Salmonella* contamination (even if properly carried out). This leads to a kind of a barn- or system-specific 'hospitalism' with a certain *Salmonella* clone, which, over time, prevails in the system in question (Van der Wolf *et al.*, 2001a; Blaha, 2008).

Even in cases where the reason for a latent *Salmonella* infection of pig herds seems to be quite obvious, it is necessary to analyse the real *Salmonella* pattern. Although in the beginning the sow farm of the study production chain was suspected to be the source of the high Salmonella load, the farm specific measures only taken in the nursery and finishing herds were suitable to reduce the infection load and the environmental contamination. In this case applying *Salmonella* reduction measures in the sow herd would not at all have led to a reduction of the *Salmonella* load of the finishing herds.

6.2. Outdoor and extensive pig husbandry is NOT at a higher risk, but the opposite is true

As the EU baseline survey in laying hens published by EFSA in 2007, the baseline survey in pigs confirmed first assumptions (Bonde and Soerensen, 2007) that outdoor pigs, always thought to be at a higher risk due to bird and rodent access and lack of biosecurity, are NOT at a higher risk of being infected with *Salmonella* than indoor pigs. In contrast, the observed prevalence in these alternative systems seems to be, as a rule, lower than that of the common intensive confinement systems. The most plausible explanation for this unexpected phenomenon is that the observed tendency of the development of barn-specific 'hospitalism' (Van der Wolf *et al.*, 2001a; Blaha, 2008) is much more pronounced in systems with a high population density and with more in-barn equipment than in the alternative husbandry systems.

7. *Salmonella* in pigs: gone tomorrow?

No, there is no doubt, *Salmonella* in pigs will not be gone tomorrow: (1) *Salmonella* is a ubiquitous part of the ecosystem and cannot be eradicated from the face of this earth; (2) there is a multitude of risk factors (see above) that are contributable to the potential introduction of *Salmonella spp.* into the food chain – this array of risk factors cannot be completely extinguished, but only be controlled.

Thus, the ultimate goal is not to eradicate Salmonella, but to keep it out of the food production chain at each level: feed, animals, transport, slaughter, processing and retail. The only way to do this is to implement quality management systems along the food chain, of which any known and feasible intervention measure capable of minimising the introduction and multiplication of Salmonella in the food chain are permanent and integral parts of the quality management system that encompasses all parts of the food chain.

References

Baggesen, D.L., Wegener, H.C., Bager, F., Stege, H. and Christensen, J. (1996). Herd prevalence of *Salmonella enterica* infections in Danish slaughter pigs determined by microbiological testing. *Preventative Veterinary Medicine* **26**: 201-213.

Beloeil, P.A., Chauvin, C., Proux, K., Madec, F., Fravalo, P. and Alioum, A. (2004) Impact of the *Salmonella* status of market-age pigs and the pre-slaughter process on *Salmonella* caecal contamination at slaughter. *Veterinary Research* **35**: 513-530.

Blaha, Th. (2008). *Salmonella* in Food Animal Production – Challenge and Chance for the Veterinary Profession. *Deutsche Tierarztliche Wochenschrifte* **56**: 906-908.

Bode, K. (2007). Serological and epidemiological investigations of the dynamics of Salmonella in swine herds for the optimization of the *Salmonella* monitoring in pigs. Doctoral Thesis, University of Hannover, Germany

Bode, K., Baier, S. and Blaha, Th. (2008). Successful *Salmonella* control in 3 finishing herds supplied by one sow herd. In: *Proceedings International Pig Veterinary Society 2008*, Durban, South Africa. 23-26 July, 2008.

Bonde, M. and Soerensen, J.T. (2007). *Salmonella* infection level in Danish indoor and outdoor production systems measured by antibodies in meat juice and faecal shedding on-farm and at slaughter. In: *Proceedings XIII International Congress in Animal Hygiene*, ISAH, Tartu, Estonia, Vol. I, pp. 729.

Dahl, J., Wingstrand, A., Nielsen, B. and Baggesen, D.L. (1997). Elimination of *Salmonella typhimurium* infection by the strategic movement of pigs. *Veterinary Record* **140**: 679-681.

Davies, R.H., McLaren, I.M. and Bedford. S. (1999). Observations on the distribution of *Salmonella* in a pig abattoir. *Veterinary Record* **145**: 655-661.

Devriese, L. (1995). Serovars of Belgian *Salmonella* Isolates Serotyped at the National-Institute-of-Veterinary-Research During the Years 1991, 1992 and 1993, Evolution Among Poultry, Pigs and Bovines 1982-1983 Doorp-Pohl - Nido, Brussels. *Vlaams Diergeneeskundig Tijdschrift* **64**: 28.

Edel, W., Van Schothorst, M., Guinee, P.A.M. and Kampelmacher, E.H. (1970). Effect of feeding pellets on the prevention and sanitation of *Salmonella* infections in fattening pigs. Zentralbladt fur Veterinärmedicin **17**: 730-738.

Edel, W., Van Schothorst, M., Guinee, P.A.M., and Kampelmacher, E.H. (1974). *Salmonella* in pigs on farms feeding pellets and on farms feeding meal. *Zentralblatt für Bakteriologie, Parasitenkunde, Infektionskrankheiten und Hygiene. Reihe A: Medizinische Mikrobiologie und Parasitologie* **226**: 314-323.

EFSA (2007). Report of the Task Force on Zoonoses Data Collection on the Analysis of the baseline study on the prevalence of Salmonella in holdings of laying hen flocks of *Gallus gallus*. The EFSA Journal **97**.

EFSA (2008). Report of the Task Force on Zoonoses Data Collection on the analysis of the baseline survey on the prevalence of *Salmonella* in slaughter pigs, in the EU, 2006-2007. Part A: Salmonella prevalence estimates. *The EFSA Journal* **135**: 1-111.

Fedorka-Cray, P.J., Hogg, A., Gray, J.T., Lorenzen, K., Velasquez, J. and Von Behren, P. (1997). Feed and feed trucks as sources of *Salmonella* contamination in pigs. *Journal of Swine Health and Production* **5**: 189-193.

Geue, L., Käsbohrer, A., Helmuth, R., Blaha, Th., Staak, C.H. and Steinbach, G. (1997). Vorkommen von Salmonellen bei Schlachtschweinen deutscher Herkunft. *Mitteilungsblatt zur Tierseuchensituation der Bundesrepublik Deutschland* **I. Halbjahr 1997**: 11-13.

Grimont, P.A.D., Grimont, F. and Bouvet, Ph. (2002). Taxonomy of the genus *Salmonella*. (20002). In: *Salmonella* in Domestic Animals (Eds. Wray, A. and Wray, C.). Oxon UK, CAB International, pp. 1-17.

Harris, I.T., Fedorka-Cray, P.J., Gray, J.T., Thomas, L.A. and Ferris, K. (1997). Prevalence of *Salmonella* organisms in pigs feed. *Journal of Animal Veterinary Medical Association* **210**: 382-385.

Hurd, H.S., Gailey, J.K., McKean, J.D. and Rostagno, M.H. (2001). Experimental rapid infection in market pigs following exposure to a *Salmonella* contaminated environment. *Berliner und Münchener tierärztliche Wochenschrift* **114**: 382-384.

Hurd, H.S., McKean, J.D., Griffith, R.W., Wesley, I.V. and Rostagno, M.H. (2002). *Salmonella enterica* infections in market pigs with and without transport and holding. *Applied and Environmental Microbiology* **68**: 2376-2381.

Letellier, A., Messier, S. and Quessy, S. (1999). Prevalence of Salmonella spp. and *Yersinia enterocolitica* in finishing pigs at Canadian abattoirs. *Journal of Food Protection* **62**: 22-25.

Lumsden, J.S. and Wilkie, B.N. (1992). Immune response of pigs to parenteral vaccination with an aromatic dependent mutant of *Salmonella typhimurium*. *Canadian Journal of Veterinary Research* **56**: 296-302.

Kamphues, J., Brünning, I., Hinrichs, M. and Verspohl, J. (2006). Investigations on counts of *Salmonella Derby* in the content of the alimentary tract in piglets 4–6 hours after experimental oral infection related to dietary influences (grinding intensity, use of potassium diformate). In: *Proceedings 19*[th] *International Pig Veterinary Society Congress*, Copenhagen, Denmark. Vol. 1, 2008.

Murray, C.J. (1991). Salmonellae in the environment. *Revue Scientifique et Technique* **10**: 765-785.

Ortmann, R. (1999). Immunisierungsversuche mit der *Salmonella Typhimurium*-Lebendvakzine Salmoporc R zur Bekämpfung von Salmonellen-Infektionen in Ferkelerzeugerbetrieben. Doctoral Thesis, Hannover, Germany

Reeves. M.W., Evins, G.M., Heiba, A.A., Plikaytis, B.D. and Farmer. J.J. (1989). Clonal nature of *Salmonella typhi* and its genetic relatedness to other salmonellae as shown by multilocus enzym electrophoresis, and proposal of *Salmonella bongori* comb. nov. *Journal of Clinical Microbiology* **27**: 313-320.

Salmon, D.E. and Smith, T. (1886). The bacterium of pigs plague. *American Monthly Microscopical Journal* **7**: 204-205.

Springer, S., Lindner, T., Steinbach, G. and Selbitz, H.J. (2001). Investigation of the efficacy of a genetically-stabile live *Salmonella typhimurium* vaccine for use in pigs. *Berliner und Münchener tierärztliche Wochenschrift* **114**: 342-345.

Swanenburg, M., Urlings, H.A.P., Keuzenkamp, D.A. and Snijders, J.M.A. (2000). *Salmonella* in the lairage of pig slaughterhouses. *Journal of Food Protection*, in press.

Van Aartsen, J. (1997). Council Directive 97/22/EC of 22 April 1997 amending Directive 92/117/EEC concerning measures for protection against specified zoonoses and specified zoonotic agents in animals and products of animal origin in order to prevent outbreaks of food-borne infections and intoxications. *Official Journal* **L113** (document 397L0022):0009-0010).

Van der Wolf, P.J. and Peperkamp N.H.M.T. (2001). *Salmonella* (sero)types and their resistance patterns isolated from pig faecal and post-mortem samples. *The Veterinary Quarterly* **23**: 175-181.

Van der Wolf, P.J., Lo Fo Wong, D.M., Wolbers, W.B., Elbers, A.R., Van der Heijden, H.M. and Van Schie, F.W. (2001a). A longitudinal study of *Salmonella enterica* infections in high- and low-seroprevalence finishing pigs herds in The Netherlands. *The Veterinary Quarterly* **23**: 116-121.

Van der Wolf, P.J., Van Schie, F.W., Elbers A.R., Engel, B., Van der Heijden, H.M. and Hunneman, W.A. (2001b). Administration of acidified drinking water to finishing pigs in order to prevent *Salmonella* infections. *The Veterinary Quarterly* **23**: 121-125.

Van der Wolf, P.J., Wolbers, W.B., Elbers, A.R.W., Van der Heijden, H.M.J.F., Koppen, J.M.C.C. and Hunneman, W.A. (2001c). Herd level husbandry factors associated with the serological *Salmonella* prevalence in finishing pig herds in The Netherlands. *Veterinary Microbiology* **78**: 205-219.

Van der Wolf, P.J., Elbers, A.R.W., Van der Heijden, H.M.J.F., van Schie, F.W., Hunneman, W.A., and Tielen, M.J.M. (2001d). *Salmonella* seroprevalence at the population and herd level in pigs in The Netherlands. *Veterinary Microbiology* **80**: 171-184.

Nutritional approaches to controlling *Salmonella*

Colm A. Moran
Alltech European Bioscience Centre, Dunboyne, Co. Meath, Ireland

1. Introduction

Salmonellosis is generally accepted to be one of the most important zoonoses transmitted by meat and eggs in the developed world. The impact of salmonellosis on human health results in a significant economic loss at the hands of this disease (Schwartz, 1990). The majority of salmonella infections arise from consumption of poultry products (Figure 1). In Europe, during 2006, a total of 5,355 food-borne outbreaks were reported (EFSA, 2007). Salmonella was responsible for 53.9% of all of these outbreaks, affected 22,705 persons and 14% were hospitalised. However, the numbers are much larger when one looks at the total confirmed individual cases of 160,949 reported from the 25 member states. In the United States the problem appears to be much greater, an estimated 1.4 million cases and more than 500 human deaths occur annually due to this pathogen (Mead *et al.*, 1999). Approximately, 95% of human *Salmonella* infections are foodborne, corresponding to approximately 30%

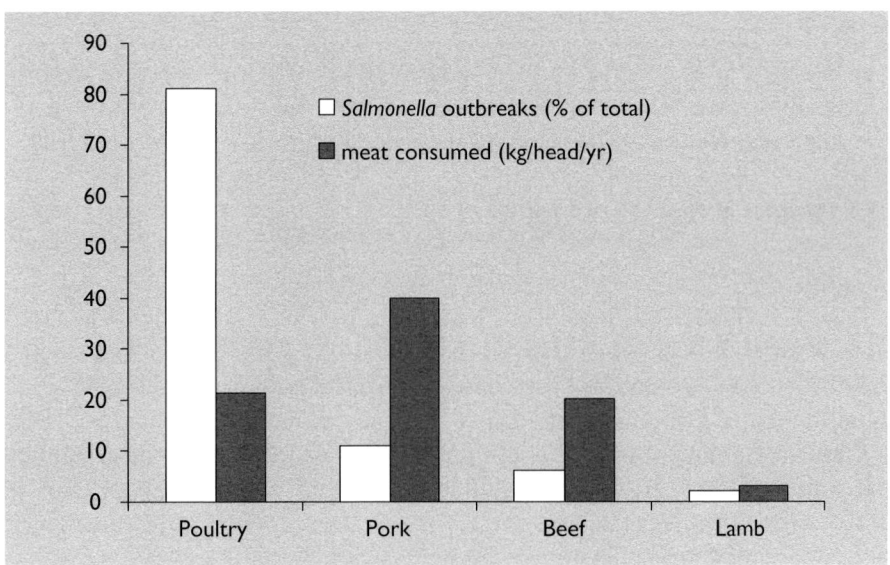

Figure 1. Human salmonellosis outbreaks as associated with animal species.

of deaths caused by foodborne infections in the United States. Weekly media reports on salmonellosis epidemics such as those identified in pork products in Ireland and Denmark and vegetables in the US have raised the awareness of the consumer to the point where legislation is being imposed by governments to control salmonella. There is little doubt that surveillance programmes that timely detect Salmonella contaminations in the entire food chain (animal feed, living animals, slaughterhouses, retail sector and restaurants) together with sanitary measures are essential to detect and prevent human *Salmonella* infections. New EU legislation will enforce control through financial penalties on producers with high levels of *Salmonella* as monitored on entry to the slaughterhouse.

Salmonellosis is caused by a gram-negative, facultative anaerobic intracellular pathogen. There are more than 2,500 serotypes of *Salmonella* which can be found in the intestinal tract of animals, birds and humans. The epidemiology of *Salmonella* is complex and can come from a variety of sources and causes sever disruption on a farm (Figure 2). *Salmonella* has the ability to survive, even under extreme conditions, for a considerable time in the environment (pH range of 4-8, low oxygen and temperature of 8-45 °C). However, in order to survive in the intestine, as well as to multiply and invade cells *Salmonella* needs to attach to receptors in the intestine. *Salmonella* adherence to host tissue is regarded as an important initial step for colonisation and pathogenesis. The majority of *Salmonella* colonise the gastrointestinal tract using specialised structures known as Type-1 fimbriae. These receptors have specificity for mannose, a sugar that is found on the surface of epithelial cells in the intestine. *In vitro* studies have shown that 80% of *Salmonella typhimurium* and 67% of *Salmonella enteritidis* express mannose sensitive fimbriae, and for *E. coli* it is over 65%.

2. *Salmonella* in pigs and poultry

2.1. Pigs

A wide variety of *Salmonella* serovars can be hosted by the pig. These serovars can be divided into two groups based on differences in pathogenesis and resulting clinical picture; Group 1: host adapted serovars primarily associated with pigs – Choleraesuis and Typhisuis and Group 2: the vast majority of the remaining serovars with broad host range including the two most common *S. typhimurium* and *S. derby*. The first group of serovars can cause severe systemic disease in pigs, however, rarely cause infections in humans or other animal species.

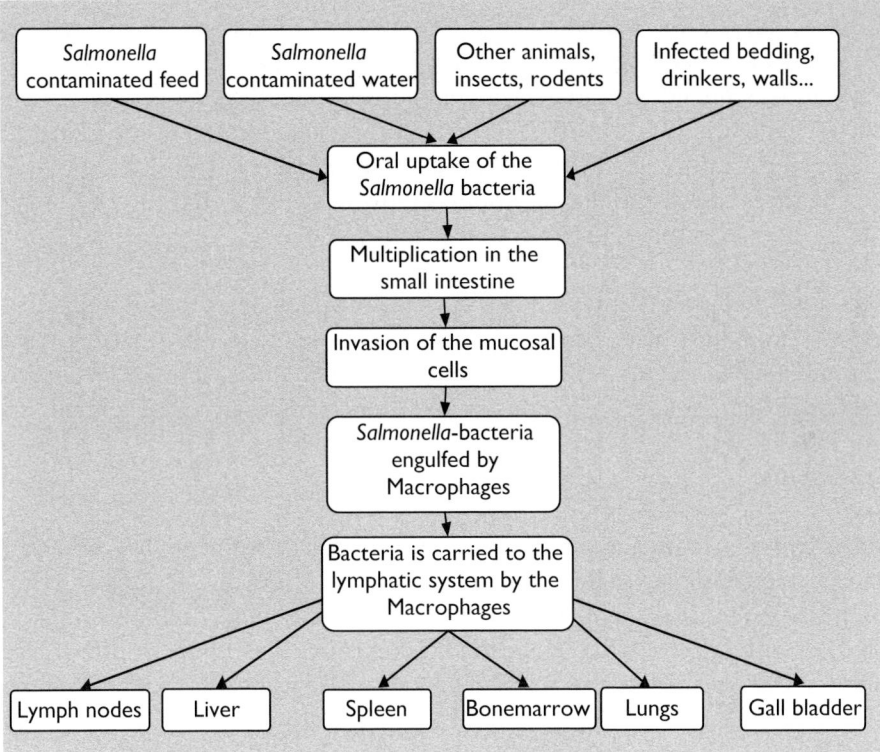

Figure 2. Salmonella *infection in livestock and poultry species.*

From the second group the serotypes of *Salmonella* such as *S. typhimurium* and *S. derby* can cause infection in pigs but also in humans. These are of public health concern. The disease is mainly manifested in pigs as enterocolitis/ diarrhoea and/or septicaemia (Kaesbohrer, 1999) and is most commonly seen in growing pigs (1st and 2nd stage weaners). Large herd size and intensive management may provide an environment conducive to *Salmonella* shedding and chronic herd infection. The excretion of considerable amounts of *Salmonella* by the first sick animals allows it to spread to others, especially those which are more susceptible (i.e. impaired immunity or liver functions). Intestinal carriage of *Salmonella* by pigs may determine the degree of carcass contamination during harvest. Consequently, European pig producers are required to conform to new legislation affecting the way they house and produce their pigs and the *Salmonella* control schemes are just one of the new standards that have to be met. The EU commission has requested that EFSA carry out an EU-wide quantitative assessment of the risk of *Salmonella* in pigs based on baseline surveys. Targets for reduction of the prevalence of *Salmonella* in herds of slaughter and breeding pigs will be established at the end of 2009 and will apply to all *Salmonella* serotypes with public health significance. However, since 2006, Regulation (EC) No 2073/2005 states that *Salmonella* must be absent in pig products, minced meat or meat products to be eaten raw. The criteria apply during the whole shelf life and products must be withdrawn from the market if they do not comply (European Commission, 2005).

This new legislation has coincided with the withdrawal of several antimicrobial growth promoters over the last number of years, some of which were known anti-*Salmonella* agents. Thus the control of *Salmonella* presents a major challenge to the modern pig industry and is one which cannot be ignored.

2.2. Poultry

As with pigs a broad range of *Salmonella* can be isolated from poultry, however, the serovars most relevant to the poultry industry are from the species *S. enterica*. These include the broad host range serovars Enerititis and Typhimurium and the host specific serovars Pullorum and Gallinarum. Depending on the serovar and the age of the bird, the Salmonella may cause infection which results in clinical disease. At the extreme end, the serovars Gallinarum and Pullorum can cause major economic damage to a poultry producer whilst other serovars may result in intestinal colonization and persistent, and albeit often intermittent, shedding in the faeces. In rare cases, mortality may be associated with invasion of these *Salmonella* outside the gastrointestinal tract. In mature chickens such as laying hens (*Gallus gallus*), the move from the intestinal tract to internal organs can be much more serious from a human food safety perspective. A classic example is that of *S. enteritidis* which can translocate to the ovary and oviduct and infect eggs prior to lay. *S. enteritidis* has caused major food safety issues in Europe and globally and is in fact the number one cause of salmonellosis in humans. In recent years, the successful vaccination of layer flocks has reduced levels of the Gallinarum, Pullorum and Enteritidis serovars in Europe.

The new EU legislation is putting the responsibility for salmonella control with the poultry industry and has set up targets for salmonella reduction in poultry (after McCartney, 2008):
- Breeding flocks tested positive for *S. typhimurium* / *S. enteritidis* must destroy non-incubated eggs. All birds, day-old chicks and hatching eggs from positive flocks must also be destroyed.

- By 2010 fresh table eggs may not be used for human consumption unless from flocks tested free of *Salmonella*. Eggs from infected layer flocks must be treated to eliminate *Salmonella* (e.g. pasteurization) prior to entering the food chain.
- By 2011 fresh poultry meat may not be used for human consumption unless tested free of *Salmonella*. Infected poultry meat must be treated to eliminate *Salmonella* prior to entering the food chain.

3. Integrated approach to *Salmonella* control on farm

Traditionally the industry has been lax in its control of *Salmonella* often relying on the final consumer to cook the product properly to eliminate the pathogen. Oftentimes the industry used in feed antibiotics to control *Salmonella* where a known flock or herd had high shedding rates of *Salmonella*. The EU has now banned the practice of including AGPs, both for growth promotion and for the reduction of *Salmonella* on farm due to increased and social pressure.

Many strategies in the past have been tried and tested, often with unsatisfactory results, for example, vaccines in pig operations. None of these *Salmonella* control strategies have been successful on their own. Therefore, the control of *Salmonella* must be considered in terms of an integrated approach combining specific nutritional technology with improved hygiene, biosecurity, and management practices which when carried out in unison will help to provide a better strategy for combating this important issue. One area that has gained significant attention in recent years is that of nutritional additives to reduce *Salmonella* colonization and / or *Salmonella* survival.

3.1. Nutritional strategies to control **Salmonella**

The development of feed additives for the control of *Salmonella* has been somewhat empirical and oftentimes the concentrations used in commercial operations is determined by cost of addition rather than the scientifically determined minimum inhibitory concentration as determined by the supplier or original scientific group. This has caused many in the poultry and pig industry to be wary of new products on the market. Clearly the product must stand up to peer review and have the necessary *in vivo* data to support the products claims. To date only a number of products have gained commercial acceptance, including:

- Short chain fatty acids – formic, acetic, propionic and butyric acids all demonstrate anti-*Salmonella* properties when added to the feed at sufficient levels. These acids are often coated or encapsulated to protect them until they reach the lower intestine where they are effective.

- Medium chain fatty acids – caproic, caprylic, capric acids. These fatty acids have been demonstrated to decrease the expression of *hilA*, a key regulator related to the invasive capacity of *Salmonella* and in addition reduce the numbers of *Salmonella in vivo* (Van Immerseel *et al.*, 2004).

- Essential oils – a range of essential oils have been shown to have bacteriostatic or bacteriocidal properties against *Salmonella*. However, many of these products are not applied at higher enough levels in practice to be effective.

- Probiotics – *Lactobacillus* spp., *Lactococcus* spp., *Bifidobacteria* spp., *Pediococcus* spp., *Enterococcus* spp., *Saccharomyces* spp. Numerous research studies have investigated the addition of live lactic acid bacteria or yeast to the diet to act in the manner of competitive exclusion following the discovery that newly hatched chickens could be protected against colonization by *Salmonella enteritidis* by dosing with a suspension of caecal contents derived from healthy adult chickens (Nurmi and Rantala, 1973). With few exceptions, the original method of application remains the best way of protecting the chick against *Salmonella* colonization. The dosing of single strains of bacteria only offers a low level of protection unless as a part of a larger program.

- Prebiotics – fructooligosaccharides and inulin. FOS is counterindicated for control of *Salmonella*. Research conducted in rats demonstrated an increased colonisation of *Salmonella* in caecal contents and enhanced translocation of *Salmonella* (Ten Bruggencate *et al.* 2004).

- Anti-infective glycans – reviewed below. Include oligomers or polymers of certain sugars, mannose, galactose, fucose, etc.

4. Glycans as the basis for anti-infective agents

Carbohydrates (glycans) have long been known to be an important dietary component, seen predominantly as energy yielding molecules and structural components of plant materials. Recently, studies have demonstrated that non-digestible carbohydrates play an important role in animal production and human health. Moreover, there is a growing recognition that non-digestible carbohydrates play a vital role in cellular metabolism, protein structure and function, cell-to-cell communication, host immunity and anti-adhesion therapy.

Mechanisms of bacterial pathogenesis have become a progressively more important subject as pathogens have become increasingly resistant to current antibiotics. Adhesion of pathogens to the epithelium surface of the gut (colonization) is the first critical stage leading to infection. It is mediated by lectins present on the surface of the infectious organism that bind to complementary carbohydrates on the surface of the host tissue. The ability to impair bacterial adhesion represents an ideal strategy to combat bacterial pathogenesis because of its importance early in the infectious process, and it's also suitable for implementation as a prophylactic to prevent infection (Cegelski *et al.*, 2008). In the case of the majority of *Salmonella*, mannose-specific lectins (Type 1 fimbriae) on the bacterial surface recognise glycoproteins (rich in mannose) on the host cell surface. Type 1 fimbriae are composed primarily of protein subunits called FimA (Kisiela *et al.*, 2006). However, their binding properties depend on the terminal protein FimH. This lectin-like adhesin, located on the tip of the fimbrial shaft, is directly responsible for bacteria binding to oligomannosidic structures adorned on many cell types including the epithelial cells of the intestinal tract. Recently the x-ray crystallography of the interaction between FimH and oligomannose-3 was elucidated (Figure 3; Wellens *et al.*, 2008). Multi-valent ligands, poly mannan and mannan oligosaccharides, are more effective than free mannose in their ability to bind to these fimbria. *In vitro* studies also have indicated differences exist in the ability of different mannose-based sugars to block pathogen attachment (Firon *et al.*, 1983). Firon and co-workers demonstrated that compounds containing both α-1,3 and α-1,6 branched mannan (as found in the outer cell wall of *S. cerevisiae*) had approximately 37.5 times the binding capacity for *E. coli* as D-mannose. In a comprehensive study of food and feed components for their adhesive properties to enteric pathogens it was found that products rich in mannan represented the most efficient binding matrices (Becker and Galletti, 2008). The conserved nature of this type of carbohydrate recognition suggests that α-1,3, α-1,6 branched mannan will have broad range efficiency in preventing type-1 fimbriae mediated adhesion and concomitant bacterial infection (Bouckaert *et al.*, 2006).

In vivo studies by Oyofo and co-workers (1989a) tested the effect of different sugars on the adherence of *Salmonella typhimurium* to epithelial cells from 1 day old chicks *in vitro* and found that mannose and methyl-α-D-mannoside were the most efficient in inhibiting adherence. They reported that mannose addition decreased the number of adherent bacterial cells to a defined intestinal surface by more than 90% when compared to a control with no carbohydrate added. In three follow-up *in vivo* studies, Oyofo and co-workers (1989a,b) observed a significant protective effect from supplementing mannose (2.5% w/v) in the drinking water of chicks for 10 d; *Salmonella*-challenged control

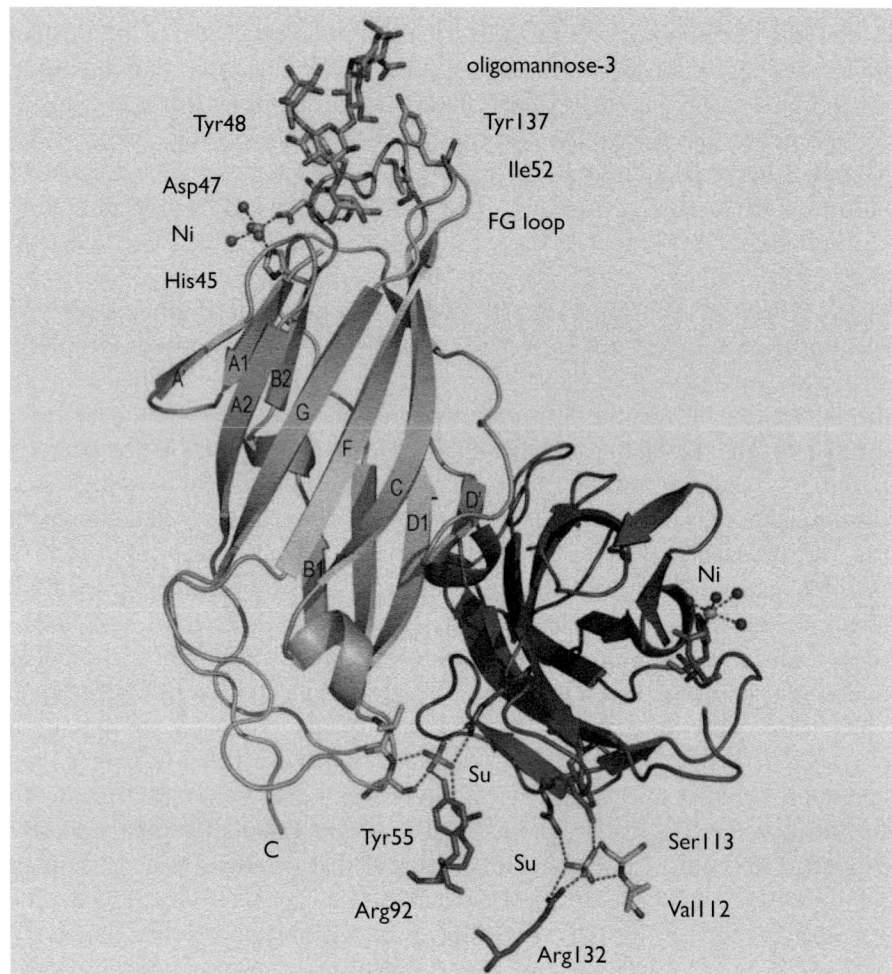

Figure 3. Crystal structure of FimH in complex with oligomannose-3 (Wellens et al., 2008).

chickens were 78, 82 and 93% colonised whereas salmonella-challenged mannose treated chickens were only 28, 21 and 43% colonised.

4.1. Structure defines function

Bio-Mos® (Alltech Inc. Nicholasville, KY) is composed of yeast call wall mannoproteins from *Saccharomyces cerevisiae* are highly glycosylated polypeptides, often 50-95% carbohydrate by weight, that form radially extending fibrillae at the outside of the cell wall (Lipke *et al.*, 1998; Kapteyn *et al.*, 1999). Many mannoproteins carry N-linked glycans with a core structure of $Man_{10\text{-}14}GlcNAc_2$-Asn structures very similar to mammalian high mannose N-glycan chains. 'Outer chains' present on N-glycans consist of 50-200 additional α-linked mannose units, with a long α-1,6-linked backbone decorated with short α-1,2 and α-1,3-linked side chains. Until recently the identification of proteins in the cell wall has been hampered by the complex nature of the cell wall structure and its relative resistance to simple digestion and extraction. A novel method was developed to tag and identify cell surface proteins using a method based on treating intact cells with a membrane-impermeable biotinylation reagent that specifically reacts with free amino groups (Casanova *et al.*, 1992). Using this method, the identity of approximately 20 cell wall-associated proteins was confirmed (Mrsa *et al.*, 1997), although following a genomic approach greater than 40 have been predicted (Smits *et al.*, 1999). Cell wall proteins can be distinguished into two distinct classes, GPI (glycosylphosphatidylinositol) and PIR (proteins with internal repeats) proteins (Kapteyn *et al.*, 1999). The GPI proteins are linked to other cell wall components through a remnant of their GPI anchor and β1,6-glucan cross-links the proteins to β1,3-glucan, an example is the α-agglutinin protein. The PIR proteins are less well understood but in contrast to the GPI proteins, are not post-translationally modified by addition of a GPI anchor, but are highly O-glycosylated (Mrsa *et al.*, 1999). The mannoproteins determine the surface of the yeast cell and are responsible for the cells antigenic behaviour. Their extraction in a functionally intact manner is relatively simple on a lab bench scale but has proved to be rather difficult in the larger batch sizes. Alltech pioneered industrial extraction procedures that are currently used to produce functional mannanoligosaccharide extracts from *Saccharomyces cerevisiae* cell wall and market under the Bio-Mos brand name.

5. Bio-Mos – mode of action

The dietary addition of Bio-Mos has been proven across species (poultry, swine, pre-ruminants and ruminants) to reduce host colonisation by *Salmonella*.

5.1. Poultry

Addition of Bio-Mos® at significantly lower dietary inclusion levels (0.4% w/w) to the mannose concentrations (2.5% w/v) used by Oyofo *et al* (1989a,b) resulted in the successful reduction of *Salmonella* and *E. coli* in the caeca of young broiler chicks (Spring *et al.*, 2000) (Figure 4). Fernandez and co-workers (2000) demonstrated a reduction in colonisation of *Salmonella enterica*

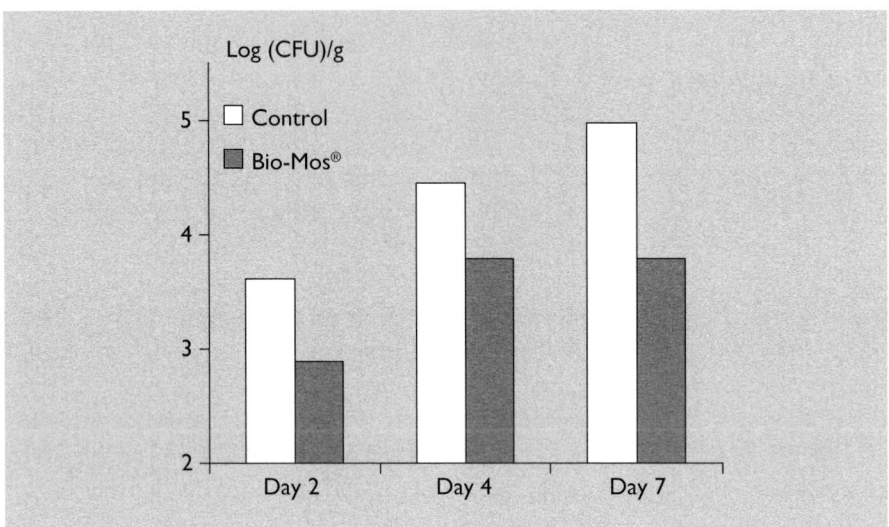

Figure 4. Effect of dietary Bio-Mos® (Alltech Inc) on caecal Salmonella typhimurium 29E concentrations of chicks at 2, 4, 7 days after salmonella challenge (Spring et al., 2000)

serovar Enteritidis (PT4) in the caeca of young broiler chicks receiving the caecal contents from hens fed MOS (2.5% w/w) through the diet. When the chicks diets were supplemented with the same MOS as given to the hens, an even greater protection was observed, as demonstrated by fewer salmonella-positive birds observed, 11/24 (46%) for MOS treatment compared with those fed mash alone 17/24 (79%). Work from the Czech Research Institute showed that birds receiving an oral gavage of *S. enteritidis* (nalidixic resistant) at the hatchery had 75% fewer cecal infections and 66% fewer infections in the Bio-Mos fed group compared to the control (negative) group (Table 1) (Sisak, 1994).

5.2. Swine

In a Danish study in 2007, the inclusion of 1 kg/T Bio-Mos in finisher diets on a commercial pig unit reduced the number of positive meat juice samples from 23.8% at the end of September to 2.8% at the end of December. In a pilot study on a UK pig farm where a strict cleaning and hygiene program had in the past failed to reduce the percentage of *Salmonella* positive meat juice samples at slaughter, the inclusion of Bio-Mos in diets at 1 kg/T was successful. The meat juice ELISA scores were reduced from 28% positive to zero over a 5 month period which in turn categorised the farm as Level 1 under the UK

Table 1. Salmonella *detection at sacrifice (birds per group) in birds receiving* Salmonella *from production unit environment and upon oral gavage with experimental strain of* Salmonella typhimurium. *Half the birds received 1kg/T Bio-Mos® through the diet (Sisak, 1994).*

Group	Wild-type *Salmonella* detected by caecal swab (birds per group)	Wild-type strain (novobiocin resistant)		Experimental strain (nalidixic acid resistant)	
Control	11/50	Caecum	38/50	Caecum	6/50
		Organs	42/50	Organs	6/50
Bio-Mos®	10/50	Caecum	9/50	Caecum	6/50
		Organs	14/50	Organs	6/50

ZAP (zoonoses action plan) scheme. These studies with Bio-Mos demonstrate that a functional carbohydrate can play an integral role in the reduction and maintenance of low *Salmonella* scores in pigs.

5.3. Selecting for antibiotic sensitivity

The emergence and persistence of antibiotic resistance in *Salmonella* continues to pose a serious health risk to human health. Despite the ban on the use of antimicrobials for growth promotion purposes there has been a continued upward trend in the identification of multiple antimicrobial-resistance determinants in different serovars including those of human health importance, *enerititis* and *typhimurium*. Therefore, there is in fact two problems when talking about *Salmonella*, a direct reduction in the numbers of *Salmonella* found in the end food product but also it would be highly desirable to reduce the resistance determinants being carried and transferred by the *Salmonella* population.

Mannan oligosaccharides have also been shown to increase the frequency of elimination of antibiotic-resistance plasmids possibly through the blocking of transmission of antibiotic resistant determinants thereby naturally allowing a bacterial population to return to antibiotic-sensitive from previously antibiotic-resistant.

The issue of reducing the prevalence and dissemination of antibiotic resistant bacteria is certainly more complex and will require a multifactorial approach. It is now clear that removing the selective pressure in the form of the antibiotics is simply one of many strategies that need to be implemented for effectively controlling the propagation and persistence of antibiotic resistance. In addition to *Salmonella* control, Bio-Mos has the potential as a nutritional technology to control the prevalence and dissemination of antibiotic resistant bacteria within the animal gastrointestinal tract. An early study by Lou and co-workers (1995) examined the potential implications of Bio-Mos on the prevalence of antibiotic resistance and the distribution of tetracycline resistance determinants in fecal bacteria from a swine herd not exposed to antimicrobials for 21 years. Fifty weanling Yorkshire pigs fed a fortified corn-soy base diet were divided into a control and a supplemented group (0.11% Bio-Mos®). Rectal swab samples were collected monthly from study initiation to market weight and 3805 lactose-positive and -negative faecal isolates were tested for resistance to 12 antimicrobial compounds (amikacin (30 mg), ampicillin (10 mg), carbenicillin (100 mg), cephalothin (30 mg), chloramphenicol (30 mg), gentamicin (10 mg), kanamycin (30 mg), nalidixic acid (30 mg), streptomycin (10 mg), sulfamethoxazole with trimethoprim (SXT) (23.75 mg/1.25 mg), sulfisoxazole (250 mg), or tetracycline (30 mg)) using standardised disk susceptibility tests. The proportion of lactose-negative isolates containing resistance to streptomycin, sulfisoxazole and tetracycline decreased ($P<0.01$) in the Bio-Mos supplemented group and increased ($P<0.01$) in the control group over time. These early observations suggested that Bio-Mos decreased the prevalence of enteric bacteria resistant to specific antimicrobial agents specifically those harbouring multiple resistance determinants in the gastrointestinal tract of pigs not subjected to antibiotic selective forces for a prolonged period of time. Although these early results strongly suggested that Bio-Mos may be an important component of a comprehensive and effective strategy for decreasing the prevalence of antibiotic resistance in bacterial populations in the gastrointestinal tract of domesticated animals it did not provide any insight into its potential mode of action.

Information encoding antibiotic resistance is most often located in plasmids and extra-chromosomal elements in bacteria. Blocking or reducing the transfer of these plasmids or extra-chromosomal elements is one way to reduce the dissemination and prevalence of antibiotic resistance encoding genes across bacterial populations. Spontaneous elimination of certain plasmids and extra-chromosomal elements is known to occur and represents another approach to controlling the proliferation of antibiotic resistant bacteria. The process of losing antibiotic resistance encoding plasmids in response to the action of certain compounds or conditions is referred to as curing, and those compounds or conditions capable of increasing the frequency of plasmid elimination are known as curing agents (Lakshimi *et al.*, 1989).

Researchers have shown Bio-Mos mediated reductions in the mean log copy number of tetracycline resistance genes in caecal contents from broiler chickens (Figure 5; Corrigan and Horgan, 2007a) and turkeys (Figure 6; Corrigan and Horgan, 2007b).

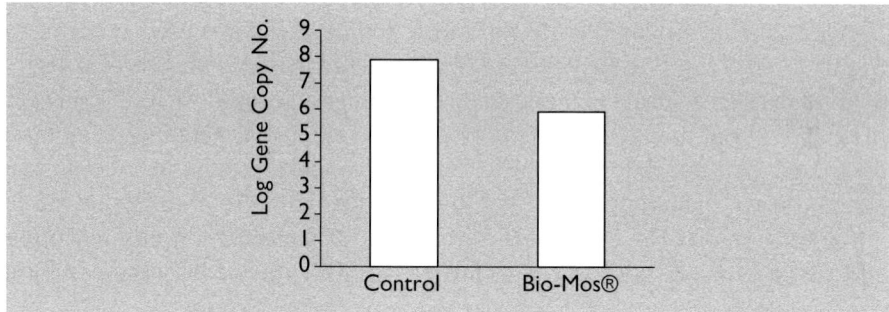

Figure 5. Effects of Bio-Mos® on levels of tetracycline resistance gene tet A in caecal contents of broiler chicks (adapted from Corrigan and Horgan, 2007a). Level of resistant gene tet A was significantly different between control and Bio-Mos® at day 21 (P<0.01).

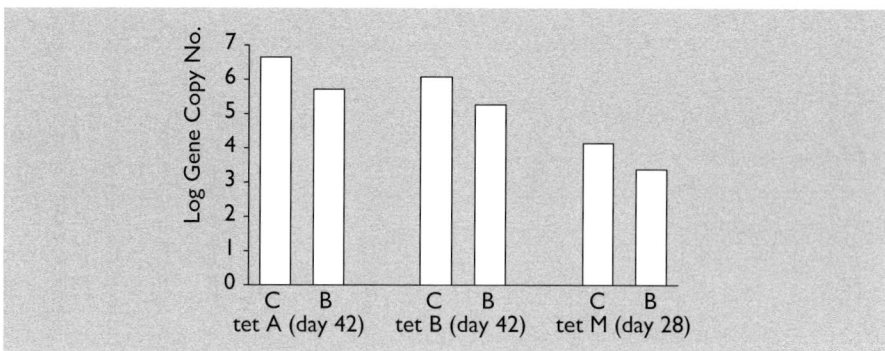

Figure 6. Effects of control (C) and Bio-Mos® (B) on levels of tetracycline resistance genes tet A, tet B and tet M in caecal contents of turkeys (adapted from Corrigan and Horgan, 2007b). Levels of resistant genes were significantly different between control and Bio-Mos® treatments at days 42 (tet A, P<0.050; and tet B, P=0.037) and 28 (tet M, P=0.026).

The ability of Bio-Mos to agglutinate bacteria bearing Type-1 fimbriae is now widely recognised (Spring *et al.*, 2000) underscoring the hypothesis that Bio-Mos mediated agglutination may interfere with bacterial conjugation thus reducing the transmission and proliferation of antibiotic resistance among bacteria. A number of studies were performed at the University of Kentucky with the objective of evaluating the ability of Bio-Mos preparations to interfere with bacterial transconjugant formation *in vitro* and in a swine faeces model (Scheuren-Portocarrero, 2004). Bio-Mos reduced (*P*<0.01) transconjugant formation by 2 log up to the initial 90 minutes of incubation when commercial *E. coli* XL1-Blue was used as donor and *E. coli* MC1000 was used as recipient in this *in vitro* broth model (Figure 7). Supplementation with Bio-Mos every 60 minutes was required to maintain transconjugant formation inhibition over the entire 3 hour incubation period (Figure 8). Bio-Mos also reduced (*P*<0.01) transconjugant formation in the swine faeces model although transconjugant formation was generally 1 to 2 log lower in the swine faeces than in the *in vitro* broth model. The doses of Bio-Mos used (0.3 and 0.5%) in these studies effectively reduced transconjugant formation even when the initial recipient concentrations were threefold greater than donor concentrations.

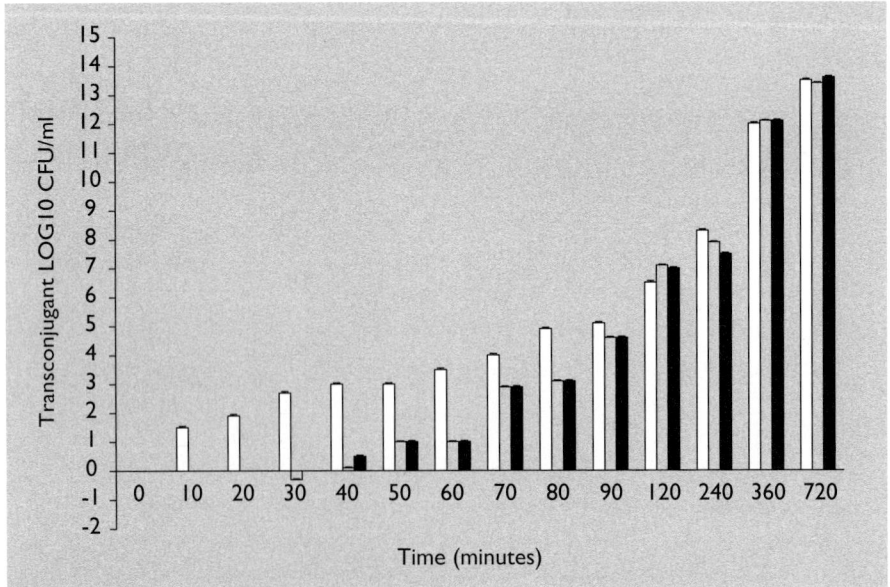

Figure 7. Transconjugant formation (+SD) during in vitro mating of E. coli *XL1-Blue (donor) and* E. coli *MC1000 (recipient) cells in the absence (empty) and presence of 0.3% MOS (grey) and 0.5% MOS (black). Significant (P<0.01) reductions in transconjugant formation were observed at 40, 60, 70, 80 and 90 minutes.*

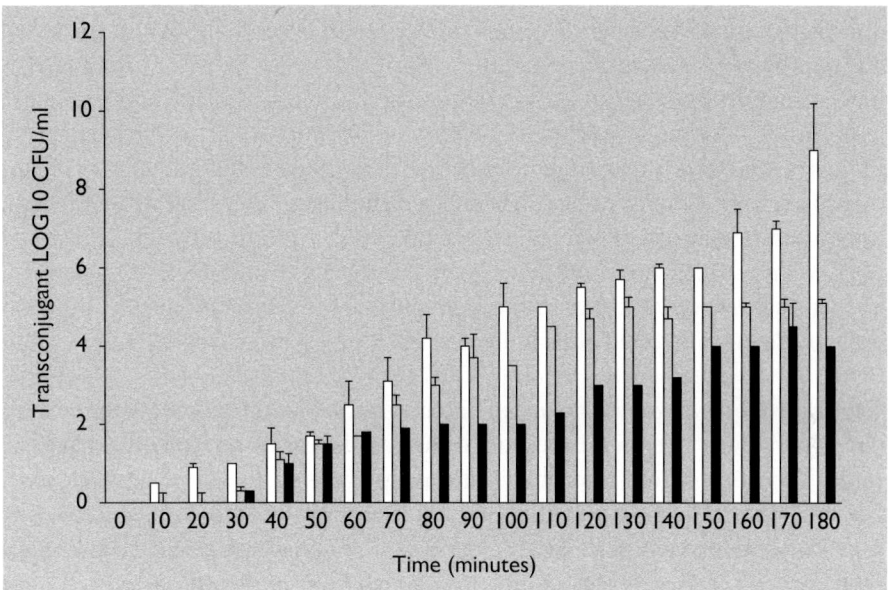

Figure 8. Transconjugant formation (+SE) during in vitro mating of E. coli XL1-Blue (donor) and E. coli MC1000 (recipient) cells in the absence of yeast cell wall-derived MOS (empty) and following the addition of yeast cell wall-derived MOS (0.5%) at the beginning (grey) and after 60 and 120 minutes of incubation (black). Significant (P<0.01) reductions in transconjugant formation were observed from 50 to 180 minutes.

Recently, Lee and co-workers (2007) have applied a stochastic compartment modelling for understanding plasmid transmission dynamics and how Bio-Mos could affect plasmid transfer in *Salmonella*. According to their mathematical model, only plasmid transfer rates influence the time required for antibiotic resistance to emerge in a newly introduced population. These researches claim that Bio-Mos may affect plasmid transfer rates by sequestering bacteria in microhabitats that exclude donors. The prevention of transference of antibiotic resistance may be due to inactivation of specific receptors on the bacterial cell wall, probably the same receptors that bind bacteria to Bio-Mos, preventing adhesion to gut mucosal cells.

6. Conclusion

Salmonellae are responsible for a number of transmissible and non-transmissible pig and poultry diseases. Public health concern about the spread of this pathogen through the food chain to humans is currently of major importance and increasingly the scientific community is concerned about the spread of antimicrobial resistance determinants within *Salmonella*. Surveillance programs have been initiated and underway within the EU, USA, Canada and Japan and a multitude of other countries. The ability of *Salmonella* to survive in a variety of environments for prolonged periods of time makes the eradication of this organism very difficult once it has entered the farm. Therefore, only through integrated management strategies that take into account biosecurity, control of pests and insects, all in/all out systems, training of staff, feed management, appropriate use of vaccines and use of specialised feed ingredients can result *Salmonella* be controlled at farm level.

The control of *Salmonella* by use of bacteriostatic or bacteriocidal compounds has traditionally been a failure due to the rapid appearance in resistance. The understanding of host-microbe relationships has led to a new generation of anti-infectives based on blocking the colonisation of the *Salmonella* and thereby preventing pathogenesis. The majority of *Salmonellae* adhere to mannose receptors in the gastrointestinal tract as a first step in the colonization of the host. The introduction of a material rich in mannose, specifically in the correct structure, allows for an oligosaccharide decoy to attach to the *Salmonella* and prevent colonization and hence infection from taking place. Bio-Mos, a mannan oligosaccharide product from *Saccharomyces cerevisiae*, is the first commercial anti-infective that can be added to feed to prevent salmonella colonization as part of an integrated strategy. Recent studies have demonstrated the ability of Bio-Mos to control the prevalence and dissemination of antibiotic resistant bacteria within the food chain.

References

Becker, P.M. and Galletti, S. (2008). Food and fed components for gut health-promoting adhesion of *E. coli* and *Salmonella enterica*. *Journal of the Science of Food and Agriculture* **88**: 2026-2035.

Bouckaert, J., MacKenzie, J. De Paz, J.L., Chipwaza, B., Choudbury, D., Zavialov, A., Mannerstedt, K., Anderson, J., Pierard, D., Wyns, L., Seeberger, P.H., Oscarson, S., De Greve, H. and Knight, S.D. (2006). The affinity of the FimH fimbrial adhesin is receptor driven and quasi-independent of *Escherichia coli* pathotypes. *Molecular Microbiology* **61**: 1556-1568.

Casanova, M., Lopezribot, J.L., Martinez, J.P. and Sentandrea, R. (1992). Characterization of cell-wall proteins from yeast and mycelial cells of *Candida-albicans* by labeling with biotin - comparison with other techniques. *Infection and Immunity* **60**: 4898-4906.

Cegelski, L., Marshall, G.R., Elridge, G.R. and Hultgren, S.J. (2008). The biology and future prospects of antivirulence therapies. *Nature Reviews Microbiology* **6**: 17-27.

Corrigan, A. and Horgan, K. (2007a). Evaluation of Bio-Mos® performance at reducing the levels of antibiotic resistant bacteria in chicken caecal contents. Poster presented at Alltech's 23rd Annual Symposium, May 20-23, 2007, Lexington, KY.

Corrigan, A. and Horgan, K. (2007b). Evaluation of Bio-Mos® performance at reducing the levels of antibiotic resistant bacteria in turkey caecal contents. Poster presented at Alltech's 23rd Annual Symposium, May 20-23, 2007, Lexington, KY.

EFSA. (2007). The Community summary report on trends and sources of zoonoses, zoonotic agents, antimicrobial resistance and foodborne outbreaks in the European Union in 2006. Available at: http://www.efsa.europa.eu/cs/BlobServer/DocumentSet/Zoon_report_2006_en.pdf

European Commission. (2005). Regulation (EC) No 2073/2005 of the European Parliament and the council of 15 November 2005 on microbiological criteria in foodstuffs. *Official Journal of the European Union* **L338**, 22.12.2005: 1; Regulation as amended by Commission Regulation (EC) No 1441/2007. *Official Journal of the European Union* **L332**, 7.12.2007: 12.

Firon, N. Ofek, I. and Sharon, N. (1983). Specificity of the surface lectins of *Escherichia coli*, *Klebsiella pneumoniae*, and *Salmonella typhimurium*. *Carbohydrate Research* **120**: 235-249.

Kaesbohrer, A. (1999). Control strategies for *Salmonella* in the pig to pork chain in the European Union. In: *Proceedings of the 3rd International Symposium on the Epidemiology and Control of Salmonella in Pork*, pp. 358-361.

Kapteyn, J.C., Van Den Ende, H. and Klis, F.M. (1999). The contribution of cell wall proteins to the organization of the yeast cell wall. *Biochimica et Biophysica Acta-General Subjects* **1426**: 373-383.

Kisiela, D., Laskowska, A., Sapeta, A., Kuczkowski, M., Wieliczko, A. and Ugorski, M. (2006). Functional characterization of the FimH adhesin from *Salmonella enterica* serovar *Enteritidis*. *Microbiology*, **152**: 1337-1346.

Lakshmi, V.V., Padma, S. and Polasa, H. (1989). Loss of plasmid linked antibiotic resistance in *Escherichia coli* on treatment with some phenolic compounds. *FEMS Microbiology Letters* **57**: 275-278.

Lee, M., Maurer, J., Soni, V. and Ewald, G. (2007). Mathematical modeling for understanding plasmid transmission dynamics and how Bio-Mos® affects plasmid transfer to *Salmonella*. Poster presented at Alltech's 23rd Annual Symposium, May 20-23, 2007, Lexington, KY.

Lipke, P.N. and Ovalle, R. (1998). Cell wall architecture in yeast: New structure and new challenges. *Journal of Bacteriology* **180**: 3735-3740.

Lou, R., Langlois, B., Dawson, K.A., Cromwell, G. and Parker, G. (1995). Effects of Bio-Mos® on prevalence of antibiotic-resistance fecal bacteria among coliforms of pigs. *Journal of Animal Science* **73** (Suppl.1): 175.

McCartney, E. (2008). Zapping *Salmonella* – facing tougher limits. *World Poultry* **7**: 2-4.

Mead, P.S., Slutsker, L., Dietz, V. McCraig, L.F., Bresee, J.S., Shapiro, C., Griffin, P.M. and Tauxe, R.V. (1999). Food-related illness and death in the United States. *Emerging Infectious Disease* **5**: 607-625.

Mrsa, V., Seisl, T., Gentzsch, M. and Tanner, W. (1997). Specific labelling of cell wall proteins by biotinylation. Identification of four covalently linked O-mannosylated proteins of *Saccharomyces cerevisiae*. *Yeast* **13**: 1145-1154.

Nurmi, E. and Rantala, M. (1973). New aspects of *Salmonella* infection in broiler production. *Nature* **241**: 210-211.

Oyofo, B.A., Droleskey, R.E., Norman, J.O., Mollenhauer, H.H., Ziprin, R.L., Corrier, D.E. and Deloach, J.R. (1989a). Inhibition by mannose of in vitro colonization of chicken small intestine by *Salmonella typhimurium*. *Poultry Science* **68**: 1351-1356.

Oyofo, B.A., Deloach, J.R., Corrier, D.E., Norman, J.O., Ziprin, R.L. and Mollenhauer, H.H. (1989b). Effect of carbohydrates on *Salmonella typhimurium* colonization in broiler-chickens. *Avian Diseases* **33**: 531-534.

Scheuren-Portocarrero, S.M. (2004). Yeast cell wall preparation as a strategy to control antibiotic resistant bacteria *in vitro* and domestic animals. Ph.D. Thesis. University of Kentucky, Lexington, KY.

Sisak, F. (1994). Stimulation of phagocytosis as assessed by luminol-enhanced chemiluminescence and response to *Salmonella* challenge of poultry fed diets containing mannanoligosaccharides. In: *Biotechnology in the Feed Industry. Proceedings of Alltech's 10th International Symposium* (Eds. Lyons, T.P. and Jacques, K.A.). Lexington, KY.

Smits, G.J., Kapteyn, J.C., Van den Ende, H. and Klis, F.M. (1999). Cell wall dynamics in yeast. *Current Opinion in Microbiology* **2**:348-352.

Spring, P., Wenk, C., Dawson, K.A. and Newman, K.E. (2000). The effects of dietary mannan oligosaccharides on cecal parameters and the concentrations of enteric bacteria in the ceca of *Salmonella*-challenged broiler chicks. *Poultry Science* **79**: 205-211.

Schwartz, K.J. (1990). Salmonellosis in mid-western swine. In: *Proceedings of the 94th Annual Meeting of the US Animal Health Association*. Denver, USA, pp. 443-449.

Ten Bruggencate, S.J.M., Bovee-Oudenhoven, I.M.J., Lettink-Wissink, M.L.J., Katan, M.B. and Van der Meer, R. (2004). Dietary fructo-oligosaccharides and inulin decrease resistance of rats to *Salmonella*: protective role of calcium. *Gut* **53**: 530-535.

Van Immerseel, F., De Buck, J., Boyen, F., Bohez, L., Pasmans, F., Volf, J. Sevcik, M., Rychlik, I., Haesebrouck, F. and Ducatelle R. (2004). Medium-chain fatty acids decrease colonization and invasion through *hilA* suppression shortly after infection of chickens with *Salmonella enterica* serovar *Enteritidis*. *Applied and Environmental Microbiology* **70**: 3582-3587.

Wellens, A., Garofalo, C., Nguyen, H., Van Gerven, N., Slättegård, R., Hernalsteens, J.-P., Wyns, L., Oscarson, S., De Greve, H., Hultgren, S. and Bouckaert, J. (2008). Intervening with urinary tract infections using anti-adhesives based on the crystal structure of the FimH–oligomannose-3 complex. *PLoS ONE* **3(4)**: e2040 doi:10.1371/journal.pone.0002040

Is immune function being traded off for optimal performance?

Brooke D. Humphrey, J. D'Amato and M. Moulds
Animal Science Department, California Polytechnic State University, 1 Grand Ave, San Luis Obispo, 93407, USA

1. Introduction

The primary responsibility of the immune system is to manage the amount and location of pathogens within the body. Increased pathogen replication or load, as well as movement of the pathogen through mucosal and tissue barriers, results in activation of the immune system. The immune system utilizes cellular and molecular defenses to reduce or eliminate pathogen load and/or remove pathogens that have breached external barriers. Activation of the immune system results in a number of behavioral and physiological changes that are associated with pathogen removal. However, many of these behavioral and physiological alterations, such as; lethargy, decreased reproduction, anorexia, decreased social interactions, increased nitrogen excretion, morbidity, mortality, limit the ability of an animal to achieve their genetic potential. In the case of poultry production, activation of the immune system can result in decreased growth and egg production. Consequently, while it is important for the animal to resist disease, the decline in animal performance is the trade-off associated with immune system activation. The ability to minimize this trade-off, that is to have animals resist disease while maintaining high levels of performance, is certainly one way of improving the efficiency and welfare of animal production systems.

Nutrition has become a tool to help achieve such an objective. Our understanding of nutrition is mature and the application of this knowledge to animal health or immunology has provided some important breakthroughs in animal production. The ability to utilize nutrition to improve disease resistance and improve animal performance requires a basic understanding of the role and regulation of nutrients in the immune system.

2. Nutrients as substrates for the immune system

2.1. Organization of the immune system and implications for nutrient use

The immune system requires nutrients at the appropriate times and in the appropriate amounts to ensure proper development, maintenance and function. Empirical performance trials are limited in their applicability for determining the nutrient needs of the immune system due to the complex organization of this physiological system. The immune system consists of numerous tissue and cell types that are responsible for the production of a vast array of effector molecules involved in pathogen killing and in the regulation of immune response (Table 1).

For example, the bursa and thymus are primary immune tissues responsible for the production of naïve B and T lymphocytes, respectively. The bursa and thymus are well-defined tissues that can be easily removed and analyzed, and their weights are often used to assess the adequacy of nutrient levels for the growth of these primary immune tissues. Since the size of the bursa and thymic lobes are indicative of the number of lymphocytes contained within this tissue,

Table 1. Nutrient consuming components and processes of the immune system.

Lymphoid tissue	Leukocytes	Molecules	Processes
Bursa	B cells	Mucus	Inflammation
Thymus	T cells	Cytokines	Acute phase protein production
Lymph node	Natural killer cells	Antimicrobial peptides	Lymphocyte development
Spleen	Monocytes / macrophages	Antibody	Lymphocyte clonal proliferation
Gut associated lymphoid tissue	Dendritic cells	Reactive oxygen species	Synthesis of new leukocytes
Cecal tonsils	Neutrophils	Acute phase proteins	Phagocytosis
Mucosal associated lymphoid tissue	Eosinophils	Reactive nitrogen species	Chemotaxis
Hardarian gland	Basophils	Antioxidants	Wound healing
Blood	Thrombocytes	Complement	
Germinal Centers			

measurements of primary immune tissue mass may provide insight into the size of the developing B and T lymphocyte populations in response to dietary nutrient concentrations. As an example, such an approach has shown that bursa and thymus mass differs with dietary arginine level (Kwak *et al.*, 1999). Additionally, relative thymus weight decreased by 30% and total thymocyte numbers decreased by 84% in response to dietary lysine levels fed below NRC recommendations (Humphrey *et al.*, 2006). In contrast, relative bursa weight and total bursacyte numbers were similar to those from chicks fed diets with adequate lysine concentrations (Humphrey *et al.*, 2006). The limitation with this approach, however, is that the applicability of these measurements to the health status of the animal is unclear. Future studies relating the size of the developing lymphocyte pools, to the function of peripheral lymphocytes are needed.

Secondary lymphoid tissues are anatomical sites where immune responses occur. These tissues provide the necessary microenvironment for antigen specific lymphocytes to proliferate in response to antigens. Consequently, secondary lymphoid tissues such as the spleen, increase in size during infections (Humphrey *et al.*, 2005) suggesting that nutrient supply to these tissues is most important during periods of immune system activation. Lymphocyte proliferation occurs in germinal centers and their total number contained within the splenic arterial tree increases 18-fold in response to vaccination (Humphrey and Klasing, unpublished results). Furthermore, the total number of splenic germinal centers is dependent upon the dietary lysine concentration (Humphrey and Klasing, unpublished results). These data indicate that the formation and amount of germinal centers, or specialized regions where immune responses occur, are impacted by substrate supply and the degree of immune system activation. The more activated the adaptive immune system, the more germinal centers will be found in secondary lymphoid tissues, such as the spleen. Therefore, presumably more nutrients will be directed toward their formation and maintenance. However, it is important to note that changes in secondary lymphoid tissue mass and structure are transient, combined with their diffuse location, this contributes to the difficulty associated with quantifying the nutrient needs for immune responses occurring in secondary lymphoid tissues.

While changes in tissue mass may provide some insight into substrate supply, a key question to consider pertains to the supply of nutrients to the leukocytes contained therein. The immune system consists of a diversity of cell types (Table 1) and at least one type of immune cell is located in every tissue in the body. Their diffuse location makes quantification difficult and assessment of substrate supply to these cells is primarily qualitative. Comparatively, lymphocytes are the most abundant leukocyte, with granulocytes and monocytes/macrophages in successive order (Klasing *et al.*, 1999). These data indicate that in a healthy, non-diseased animal, lymphocytes, based upon their abundance, are likely to be the largest nutrient consumer among leukocytes. On a body weight basis lymphocytes, granulocytes and monocytes/macrophages constitute 0.24%, 0.14% and 0.03% of a chick's body weight (Klasing *et al.*, 1999). These data indicate that the pools of leukocytes represent a small proportion of the animal's body weight and therefore, the total amount of an animal's nutrient requirement consumed by leukocytes is likely to be small.

One major determinant of the nutrient needs for leukocytes pertains to their activation state. Activated leukocytes increase their metabolic capacity in order to produce the necessary effector molecules involved in host defense. In the case of lymphocytes, their activation response involves cell proliferation and antibody production. Consequently, energy and amino acid substrates are important nutrients for these cell types (Ardawi *et al.*, 1985; Humphrey *et al.*, 2004). Cells of the innate immune system do not proliferate when activated but rather produce killing compounds, such as reactive oxygen and nitrogen species. Consequently, substrates that are metabolized to produce reducing equivalents or substrates that are products of effector molecules are of particular importance to these cell types. Examples of such nutrients are glucose, glutamine and arginine (Newsholme *et al.*, 1996; Newsholme *et al.*, 1986). Similar to secondary lymphoid tissues, the change in nutrient need for immune cells is dependent upon the activation state. Furthermore, alterations in nutrient needs are transient and since pathogens are located at specific sites within the body, the activation of immune cells is selective and often focused on a particular region, such as the intestine or lungs. Consequently, nutrient needs for immune cells are dynamic and in part dependent upon the immune cell types present and recruited to the site of infection.

2.2. Estimates of nutrient costs for immunity

Despite the challenges associated with empirical approaches to determining the nutrient needs of the immune system, factorial approaches have been utilized to sum the nutrient costs associated with developing, maintaining and using the immune system. Using lysine as a reference nutrient, Klasing and Calvert (1999) determined that the development, maintenance and use of the immune system accounts for: 1-2%, 0.5-2% and 7-10% of lysine intake respectively, in a healthy growing chick. The increase in lysine utilization during periods of infection is largely due to an increase in liver size associated with increased synthesis of acute phase proteins (Barnes *et al.*, 2002). The diversion of lysine from growth to immunity accounts for approximately 60% of the reduced growth rate in infected animals, with the remainder being attributed to anorexia associated with the early inflammatory response to infection or acute phase response (Klasing *et al.*, 1999). Similar factorial summations for other nutrients have yet to be determined and depending upon the nutrient, the immune system may account for a higher or lower amount of intake. Regardless of the absolute amount of nutrient consumed, the immune system must compete with other tissues such as skeletal muscle and reproductive tissue, for nutrients. This requires an understanding of the immune system's priority for or ability to acquire, nutrients.

2.3. Priority of the immune system for nutrients

Differences in a tissue's ability to compete for nutrients, creates a nutrient priority framework that Hammond originally proposed to explain nutrient partitioning between tissues (Hammond, 1944). According to this nutrient partitioning scheme, if more nutrients were provided in the diet, then more of those nutrients would be utilized by lower priority tissues, such as adipose. In Figure 1., on the right side are proposed sites where the immune system and leukocyte populations are thought to reside within the nutrient priority framework during both activated and inactivated states. It has been speculated that the position of the immune system within this nutrient priority scheme and the immune system's priority for nutrients, is likely to differ between activated and inactivated states, as well as between cells of the innate and adaptive immune systems (Humphrey *et al.*, 2002).

An alteration in resource availability triggers metabolic adaptation and reallocation of existing nutrient stores to support different life-traits. Consequently, understanding how nutrients are utilized by the immune system has important implications for animal health. Unfortunately, the nutrient requirements of the immune system are not known and this prohibits precise formulation of diets that are optimal for immunity. The immune system must be able to acquire nutrients at the appropriate times and amounts to ensure protective immunity. One mechanism of ensuring nutrient supply to meet this metabolic need is through unique changes in the types and amounts of nutrient transporter expression.

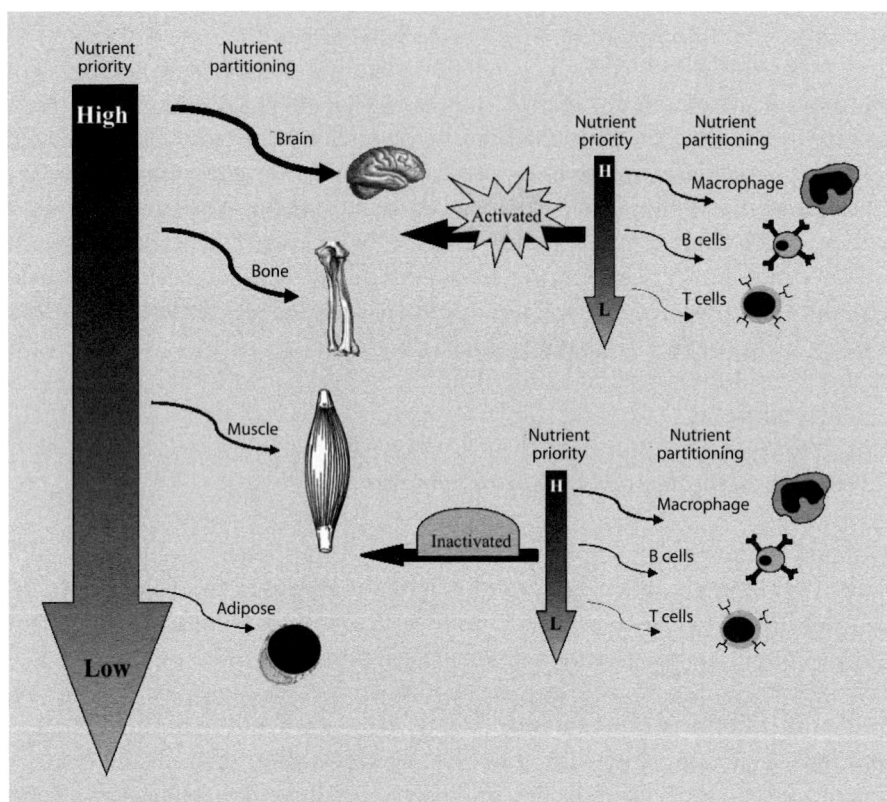

Figure 1. Priority of nutrient use by various tissues based upon their metabolic rate (adapted from Hammond, 1994).

Nutrient transporters mediate substrate specific uptake across the plasma membrane. Nutrient transporter families have multiple isoforms that differ in substrate affinity (K_m), maximum rate of transport (V_{max}) and tissue distribution. The substrate specificity of nutrient transporters, combined with multiple isoforms differing in transport kinetics, provides a robust repertoire for tissues to modulate their nutrient priority status. Three mechanisms can be used by the immune system to increase nutrient uptake to fuel the increased nutrient demand that occurs following engagement of a pathogen. First, increased nutrient acquisition is accomplished by increasing the total number of nutrient transporters on the plasma membrane, thereby increasing V_{max}. Second, increased nutrient consumption is met through the induction of a high affinity isoform of the nutrient transporter that was previously expressed at a low level, resulting in a higher K_m. Finally, the increased demand is met by a combination of the above two mechanisms. Comparing functional categories of transporters can be used to determine the priority of tissues for a particular nutrient. For example, those tissues expressing high affinity isoforms have a high priority for a particular nutrient compared to those tissues expressing a low affinity isoform. Furthermore, most high affinity isoforms transport their substrate with maximum velocity. We previously examined tissue priority for lysine and arginine by comparing functional classes of their transporters.

Lysine and arginine are accumulated into tissues by cationic amino acid transporters (CAT). To determine the ability of the immune system to acquire lysine and arginine, the expression levels of CAT isoforms in primary immune tissues were determined and compared to CAT levels from skeletal muscle, heart, liver and spleen. In two-week old broilers, the expression of high affinity CAT isoforms is approximately four-fold greater in the bursa than in the pectoralis and thymus (Humphrey *et al.*, 2004), indicating that the thymus may be most susceptible to lysine and arginine availability. Indeed, when dietary lysine levels were reduced, the transcription of genes involved in lysine and arginine uptake was reduced, indicating a decreased ability to acquire lysine and arginine (Humphrey *et al.*, 2006). Reduced expression of genes involved in lysine and arginine uptake was correlated to reduced tissue growth, indicating that less of these essential amino acids were partitioned to this tissue when their levels in the diet were reduced.

Unlike their primary immune tissues, resident lymphocytes exclusively utilize high affinity CAT isoforms to accumulate lysine and arginine (Humphrey *et al.*, 2006). Assuming that chicken CATs have similar transport kinetics as mammals, the plasma lysine and arginine concentration is above the K_m for the high affinity CAT isoforms. As a result, all CAT isoforms in thymocytes and bursacytes are transporting lysine and arginine at maximum velocity. This suggests that developing lymphocytes should be receiving an adequate supply of lysine and arginine when adequate levels of these amino acids are included in the diet. During a period of lysine deficiency, it is the reduction in food intake and not the limited lysine supply that alters thymocyte and bursacyte CAT levels (Humphrey *et al.*, 2006). Bursacytes increase high affinity CAT expression and thymocytes decrease high affinity CAT expression. This indicates that the ability of developing lymphocytes to alter their capacity to obtain lysine may be more dependent upon food intake than the dietary level of lysine.

In addition to CATs, developing chicken lymphocytes also express nutrient transporters for glutamine, fatty acids and glucose (Rudrappa *et al.*, 2007).

These latter nutrient transporters are important for the acquisition of energy substrates to support the high rates of lymphocyte proliferation. Increased expression of glutamine and glucose transporters presumably account for increased utilization of these energy substrates by these cells. Indeed, increased glucose transporter expression is associated with increased glucose uptake in developing chicken lymphocytes (Rudrappa *et al.*, 2007).

Macrophage activation results in increased utilization of nutrients to synthesize killing compounds and this increased demand is supported by increased expression of nutrient transporters. For example, activated macrophages increase expression of arginine transporters to supply these cells with sufficient arginine for increased nitric oxide (NO) production (Nicholson *et al.*, 2001). In fact, mammalian macrophages cultured in vitro increase arginine transport in response to increased arginine concentrations. In turn, increased arginine uptake results in greater nitric oxide production. However, in vivo, plasma arginine levels remain unchanged during infection (Humphrey *et al.*, 2005), indicating that this amino acid may not be limiting for macrophage function.

In addition to nutrient acquisition, metabolic enzymes play an important role in directing nutrient flux through specific metabolic pathways. The coordinated up-regulation of both nutrient transporters with specific metabolic enzymes allows the immune system to coordinate nutrient utilization. For example, developing lymphocytes require energy generation to support the high rates of cell proliferation in primary immune tissues. In chickens, developing B and T lymphocytes alter the activity of enzymes involved in the oxidation of amino acid, glucose and lipid (Rudrappa *et al.*, 2007) and this increase is coordinated with periods of heightened proliferation.

2.4. Periods of critical substrate need: Adaptive immunity

The adaptive immune system consists of B and T lymphocytes that are capable of recognizing and responding to specific foreign pathogens. Lymphocytes generate antigen specificity through the development of their antigen receptors within primary immune tissues and in birds this process begins *in ovo* and continues for several weeks posthatch. Compared to antigen-dependent proliferation, the proliferative events occurring during lymphocyte proliferation are thought to be larger since proliferation is for all lymphocytes rather than just for those that are specific to a particular antigen (Humphrey, 2005).

In the bursa, B cell development begins on embryonic day (E) 15 and progenitor cells proliferate 1,000-fold to produce 2-4x10^7 B lymphocytes by hatch (Pink, 1986; Pink *et al.*, 1985). For several weeks posthatch the B lymphocytes within the bursa double every 10 hours resulting in the production of approximately 10^7-10^8 B lymphocytes per day (Pink, 1986). This large daily production of B lymphocytes is attributed to the bursa producing 95% of the peripheral blood B lymphocytes in a four-week old chicken (Paramithiotis *et al.*, 1994). By six weeks posthatch, bursa size and B lymphocyte production is maximum (Pink, 1986). Beginning around 12 weeks the bursa begins to involute and no longer functions as a primary immune tissue for B lymphocyte development (Funk *et al.*, 2003). Therefore, the proposed period of critical nutrient need for the development of B lymphocytes spans from E15 through to six weeks posthatch due to the importance of B lymphocyte proliferation and gene rearrangement events associated with the development of humoral immunity to pathogens.

In the thymus, progenitor T lymphocytes colonize the thymus in three waves during embryogenesis, the first of which occurs on E6.5. Colonization of the thymus by progenitor T lymphocytes lasts one to two days and then these cells begin to rapidly proliferate. Two weeks are required to generate the maximum number of T lymphocyte progeny and these cells begin to emigrate to the periphery during the third week post-colonization (Cooper *et al.*, 1991). Consequently, most of the T lymphocyte development occurs *in ovo*. Proliferation of the third wave of T lymphocyte progenitors continues through the first week posthatch and progeny complete emigration by the third week posthatch. Therefore, the proposed period of critical nutrient need for developing T lymphocytes spans from E6.5 to three weeks posthatch.

The different periods of critical nutrient need between B and T lymphocytes is primarily related to differences in their development. The majority of T lymphocyte development occurring *in ovo* suggests that embryonic nutrition may contribute more than posthatch nutrition to the development of T lymphocytes. Furthermore, the thymus produces a sufficient quantity and diversity of T lymphocytes for cell mediated immunity during embryogenesis, since removing the thymus at hatch does not eliminate cell mediated immunity in later life (Chen *et al.*, 1989). The majority of B lymphocyte development occurring posthatch suggests that posthatch nutrition may contribute more than *in ovo* nutrition to the development of B lymphocytes. The majority of B lymphocyte production by the bursa occurs posthatch and removal of the bursa at hatch permanently impairs humoral immunity (Cooper *et al.*, 1969).

Consequently, it is crucial that developing B and T lymphocyte receive an uninterrupted supply of nutrients.

2.5. Periods of critical nutrient need: innate immunity

The innate immune system is responsible for protecting the host during the initial stages of infection. The innate immune system consists of constitutive defenses, such as phagocytic cells, epithelial barriers, secretions and preformed protective molecules that act as barriers to pathogen invasion. Unlike lymphocytes, cells of the innate immune system do not undergo proliferative events within primary immune tissues and all dividing cells are functional. This suggests that the development of the innate immune system may not be a critical period of nutrient need, especially compared to that of developing lymphocytes. Upon activation, however, the innate immune system becomes anabolic, resulting in the synthesis and secretion of protective proteins and killing compounds that function to eradicate the invading pathogen. The synthesis of these protective proteins, such as acute phase proteins by the liver, and killing compounds, such as nitric oxide by macrophages, require additional nutrients (Suffredini *et al.*, 1999). Therefore, activation of the innate immune system is proposed as a critical period of nutrient need. The innate immune system has no immunological memory to antigens, so repeat antigen exposure results in an innate immune response similar in magnitude to the initial response. Consequently, the nutrient need for the activated innate immune system will be similar for each encounter with the same antigen.

2.6. Substrates of importance to the immune system

The immune system has specific energy and amino acid substrates that are preferred for supporting the anabolic events associated with activation. These substrates are important for fueling the proliferation of lymphocytes and the synthesis of protective factors associated with host defense.

Energy metabolism is of particular importance to lymphocytes since their development and activation require cell proliferation. Leukocytes primarily utilize glucose and glutamine as an energy source (Ardawi *et al.*, 1985). Glucose is the fuel of choice for lymphocytes and in mammals glucose is actually an essential nutrient for lymphocytes (Greiner *et al.*, 1994). Glucose is also important for generating reducing equivalents through the pentose phosphate pathway. These reducing equivalents are essential for producing killing compounds involved in the macrophage respiratory burst (Newsholme *et al.*, 1996). Second to glucose, glutamine is also a major energy substrate for leukocytes. Like glucose, glutamine metabolism can also generate reducing equivalents necessary for the production of reactive oxygen species (Newsholme, 2001). Glutamine conversion to glutamate may also aid in the transport of amino acids since glutamate is a substrate for many amino acid transport exchange systems (Aledo, 2004).

Arginine can be metabolized to produce nitric oxide involved in inflammatory responses and polyamines involved in wound healing. In mammals, arginine plays an integral role in the development of B lymphocytes (De Jonge *et al.*, 2002) and also regulates the signaling ability of T lymphocytes (Rodriguez *et al.*, 2002). Since arginine can be synthesized in ureotelic species, immune cells, particularly macrophages, are capable of synthesizing and recycling arginine for use in nitric oxide production. In uricotelic species and strict ureotelic carnivores that are incapable of arginine synthesis, this amino acid is not only essential but cannot be recycled (Sung *et al.*, 1991). Consequently, this amino acid is of particular importance in these species during periods of infection since increased endogenous synthesis of arginine cannot compensate for increased utilization of this amino acid by the immune system.

A considerable amount of research has been conducted in chickens examining the effect of arginine on the immune system (Kidd *et al.*, 2001; Kwak *et al.*, 1999; Takahashi *et al.*, 1999) and the provision of this nutrient to the immune system appears most important for activation. For example, avian macrophage activation results in increased synthesis of nitric oxide. Since arginine is required for nitric oxide production, it is not surprising that these cells increased the expression of genes involved in arginine uptake and decrease the expression of genes involved in arginine export (Moulds and Humphrey, unpublished). The net result is increased arginine intake to help supply the increased flux of this amino acid through the nitric oxide synthesis pathway.

Glutamine has long been recognized as an important metabolic fuel for the immune system (Ardawi *et al.*, 1983). Glutamine is primarily metabolized to generate energy for leukocytes, as well as reducing equivalents for synthesizing reactive oxygen intermediates (Newsholme, 2001). Glutamine can also be metabolized to arginine in murine macrophages for NO synthesis (Murphy *et al.*, 1998). Although glutamine can be synthesized endogenously, differences in glutamine metabolism between uricotelic and ureotelic species may have implications on glutamine use by cells of the immune system between animals with these nitrogen excretion strategies.

Cysteine can be metabolized to produce glutathione (GSH). GSH is one of the major intracellular antioxidants and its production is regulated by the availability of cysteine. GSH production increases during periods of inflammation (Malmezat *et al.*, 2000) and consequently a greater proportion of cysteine metabolism is directed toward GSH synthesis (Malmezat *et al.*, 1998). GSH plays an important role in leukocyte function and these cells have a strong ability to obtain cysteine (Droge *et al.*, 1991). Increased cysteine utilization for GSH synthesis, results in taurine production and inflammatory responses, resulting in increases taurine levels in the liver and kidney. However, taurine levels are reduced in the gastrointestinal tract (Malmezat *et al.*, 1998) and this reduction may have implications on bile salt formation and lipid absorption in carnivores.

2.7. Supplying nutrients to the immune system: push versus pull

When considering diet modifications to help feed the immune system and optimize animal health, it is important to consider the concept of 'push versus pull', in regards to nutrient partitioning to the immune system. A common approach aimed at increasing the activity of the immune system, is to include more of a particular nutrient that is suspected to be in limited supply in the diet or during a particular physiological state. This 'push' approach to feeding the immune system assumes that more is better and is fundamentally based upon the idea of tissue competition for nutrients, such as skeletal muscle versus immune tissue (Figure 1). The 'push' approach to feeding the immune system assumes that providing more nutrients in the diet will increase their utilization by the immune system. However, simply increasing the supply of a particular nutrient does not necessary directly translate to increased utilization. For example, increased plasma fatty acid levels will result in increased passive diffusion into tissues, yet this does not translate into increased fatty acid esterification or oxidation. These events are also regulated by signaling systems that act independent of the nutrient supply per se. Consequently, the 'push' approach to feeding the immune system is based upon nutrient supply alone and does not consider the controls that couple nutrient supply with nutrient demand. With regard to feeding the immune system, it is important to understand how these cell types and tissues coordinate nutrient utilization, as well as to understand the periods when nutrient utilization is of critical importance to the immune system (discussed above).

The 'pull' approach to feeding the immune system involves modulating the coordinated adaptations that direct nutrient partitioning toward immune function. These adaptations are complex and regulated by signals, often cytokines, to help coordinate nutrient supply with nutrient demand across all tissues and within the animal (Humphrey *et al.*, 2004). The coordinated adaptations of nutrient utilization throughout the body by signals from the immune system help to ensure that nutrient supply meets the nutritional demands associated with immune function. To ensure nutrient acquisition during these times, immune cells can utilize several strategies to partition more nutrients toward immunity.

Coordination of nutrient partitioning to the immune system is further illustrated by reciprocal regulation of nutrient metabolism in skeletal muscle and immune tissue during periods of infection. In general, tissues engaged in host defense require greater amounts of amino acid substrate to synthesize protective factors and to support leukocyte proliferation (Klasing *et al.*, 1984). Those tissues not directly involved in immune defense are either unaffected or increase their rates of protein degradation (Klasing *et al.*, 1984). Skeletal muscle is the largest labile pool of amino acids and infection increases protein degradation and release of amino acids into plasma (Hentges *et al.*, 1984; Tian *et al.*, 1989; Tian *et al.*, 1989). The release of amino acids from skeletal muscle plays an important role in supplying amino acid substrate for the synthesis of hepatic acute phase proteins involved in host defense (Barnes *et al.*, 2002). This coordination of altered nutrient partitioning during infection is related to changes in amino acid transporter expression, with elevated levels in the skeletal muscle to allow for amino acid liberation and elevated levels in the liver to allow for increased amino acid absorption for synthesis of acute phase proteins (Humphrey *et al.*, 2005).

Coordination of nutrient partitioning to the immune system is achieved in large part through the actions of cytokines. Leukocytes display many cytokine receptors but other cell types have a much more limited expression pattern. This difference allows cytokines to act selectively upon cells of the immune system to increase their nutrient acquisition. Furthermore, the selective action of cytokines can affect nutrient acquisition by specific leukocyte populations. For example, activated T lymphocytes produce interleukin-2 (IL-2) that acts in an autocrine and paracrine manner to increase T lymphocyte proliferation. IL-2 also increases T lymphocyte glucose transporter-1 (GLUT-1) protein to provide energy for proliferation. During lymphocyte development, IL-3 also increases lymphocyte nutrient transporters for glucose, amino acids, lipids and metals (Edinger *et al.*, 2002) and IL-7 maintains the metabolic activity of naïve T lymphocytes (Rathmell *et al.*, 2001).

Activation of lymphocyte antigen receptors is also a key regulatory event that signals for increased nutrient utilization (Fox *et al.*, 2005). Antigen receptor activation triggers the lymphocyte to undergo clonal expansion, a proliferative event, to allow for the generation of antigen specific lymphocytes. Signaling pathways emanating from the antigen receptor aid in the coordination of lymphocyte metabolism to meet the energetic demands of activation (Frauwirth *et al.*, 2004).

2.8. Supplying substrates to the immune system: the application

When formulating diets that are optimum for the immune system, it is important to consider the type of nutrients being offered to the immune system, i.e. supply, and to what specific aspect of the immune system that they are intended for, i.e. demand. Considering supply without demand can result in either no impact on immune function or even decreased overall animal health, as evidenced by the severity of *E. coli* infection in iron supplemented newborn pigs (Kadis *et al.*, 1984). Rather, nutritional approaches to enhance immune function should focus on supplying the nutrients at the appropriate times and in the appropriate amounts that complement the 'pull' associated with increased nutrient partitioning for immune function.

3. Nutrients regulate the type of immune response

The immune response is organized to protect the host from intracellular and extracellular pathogens. Dependent upon the location of the pathogen, the immune system coordinates a specific response that is tailored towards eliminating that pathogen (Goldsby *et al.*, 2003). For example, extracellular pathogens such as *E. coli*, elicit an innate inflammatory response as well as an adaptive antibody response, whereas intracellular pathogens such as viruses, elicit a cell mediated immune response. The release of cytokines by immune cells in response to pathogen recognition provides instruction for coordinating the appropriate type of immune response to eliminate the extracellular or intracellular pathogen. T helper 1 and 2 (Th1 and Th2, respectively) lymphocyte populations and macrophages are important cell types involved in the production of cytokines in response to pathogen; therefore, these cells function to direct the type of immune response elicited to a particular pathogen.

Nutrients can regulate the type of immune response and this is in large part attributed to their ability to alter cell communication pathways and/or gene expression. Altered cell communication pathways and gene expression provide instruction on how, when and for how long to respond which are all important criteria for mounting an immune response. Examples of immunomodulatory nutrients include vitamin A, E and polyunsaturated fatty acids (PUFAs). Table 2 provides a summary of some of the published reports that have examined the effect of these nutrients on immune function in broilers, layers and turkeys. The mammalian literature is much more replete with reports on the immunomodulatory properties of nutrients and helps to provide basic information that can aid in the understanding of how these nutrients may modulate the immune response in poultry.

The impact of these nutrients on immune function is dependent upon nutrient dose, type of pathogen challenge, dose of pathogen challenge and immune parameter measured (Table 2). Often these immunomodulatory properties are realized when these nutrients are included in the diet above NRC recommendations. However, the effect of some of these immunomodulatory properties is curvilinear and higher dietary concentrations can have adverse effects. The majority of studies have also utilized broilers and there is limited information on how these and other nutrients may modulate immune function in layers and broilers. Given the divergence in immune response between broilers and layers (Leshchinsky *et al.*, 2001), applying results obtained in broilers to layers may not be justified. More research examining the impact of nutrition on the immune response of turkeys is needed.

As nutrients continue to be examined for their immunomodulatory properties, it is important to carefully consider the selection criteria utilized to evaluate their effectiveness or response (Fulton, 2004). Specifically, it will be important for future studies in this area to select assays that are accurate in their estimation and are relevant and/or related to disease resistance (Cunningham-Rundles, 1998). As mentioned previously, it is not clear how changes in relative immune organ weights relate to disease resistance. Furthermore, it is assumed that lymphocyte proliferation *in vitro* to a common mitogen is similar to the proliferation of any signal that triggers proliferation of lymphocytes *in vivo*, such as antigens. Additionally, *in vitro* assays commonly use standardized cell numbers as well as isolated cell populations, which does not reflect the cellular environment *in vivo*. This requires careful selection of measurements of immune response and the development of new methodology in this area, which fit these criteria, may be constructive. These are important steps toward understanding the ability of nutrients to modulate the immune system in a relevant context.

Table 2. Effect of nutrients on immune function in broilers, layers and turkeys[1].

Nutrient	Species	Amount	Challenge	Assay	Effect	Reference
Vitamin A	Broiler	8000 IU/kg	Eimeria acervulina	IEL	↑ CD8+ IEL	Dalloul et al., 2002
	Broiler	8000 IU/kg	Eimeria acervulina	Titre	↑ titre 6 d post infection	Dalloul et al., 2003
	Broiler	0.85, 1 mg/kg	E. coli	Titre	1 g/kg ↑ mortality & morbidity 1 g/kg ↓ titre	Friedman et al., 1991
	Broiler	6.6, 13.2 mg/kg	β-casein	T cell proliferation	optimal @ 6.6 mg/kg ↓ 13.2 mg/kg	Halevy et al., 1994
	Broiler	1500, 15000 IU/kg	Mitogen, NDV vaccine	Proliferation	1500 IU ↓ proliferation to mitogens	Lessard et al., 1997
	Layer	4000,12000, 24000 IU/kg	NDV vaccine	Titre, T cell counts	No effect	Coskun et al., 1998
	Layer	3000, 9000 IU/kg	LoSota vaccine, NDV vaccine	Agglutination	No effect	Lin et al., 2002
	Turkey	6.6, 13.2 µg/g	NDV, pox vaccines, Mitogen	Titre, Proliferation	6 µg/g maximum T cell proliferation 13.2 µg/g ↑ titre	Sklan et al., 1995
Vitamin E	Broiler	25-5000 mg/kg	Tetanus toxoid	Intestinal IgA	↑ IgA @ 5000 mg/kg	Muir et al., 2002
	Broiler	10.2-210.2 IU/kg	IBV antigen, SRBC	Titre	↑ IBV titre between 10-35 IU/kg ↑ SRBC titre @ 50 IU/kg	Leshchinsky et al., 2001
	Broiler	13-200 ppm	Eimeria maxima	Oocyte shedding	200 ppm ↓ oocyte shedding	Allen et al., 2002
	Broiler	0, 30, 200 IU/kg	LPS	T cell profiles, Cytokines	↑ CD4-CD8+ ↓ CD4+/CD8+ @ 30 IU/kg 200 IU/kg ↓ TGF-β expression	Leshchinsky et al., 2003

Table 2. Continued.

Nutrient	Species	Amount	Challenge	Assay	Effect	Reference
Vitamin E (continued)	Broiler	0.03%	NDV vaccine	Titre, Proliferation	↑ proliferation ↑ titre of offspring at d 1 & d 7	Haq et al., 1996
	Broiler	0, 10, 30, 150 mg/kg	E. coli, NDV, pox vaccine	Titre	High dose ↓ titre	Friedman et al., 1998
	Turkey	14 IU/lb/d	E. coli	WBC counts	↑ heterophil/lymphocyte ratio	Huff et al., 2004
	Turkey	0, 50, 150 mg/kg	E. coli, NDV vaccine, Turkey pox vaccine	Titre	High dose ↓ titre	Friedman et al., 1998
n-3 & n-6 PUFA	Broiler	1, 1.5, 2% Fish oil	IBV Vaccine, LPS	Cytokine	↓ IL-1 activity	Korver et al., 1997
	Broiler	5% corn, linseed, menhaden or tallow	LPS	Cytokines	↑ IL-2	Sijben et al., 2001
	Broiler	% corn, linseed, menhaden or tallow	LPS	Cytokines	↑ IFN-γ	Sijben et al., 2003
	Broiler	2.64-6.24 % LA, 0.09-2.43% LNA	KLH, M. butyricum	Titre, Proliferation	↑ titre LNA ↑ proliferation	Sijben et al., 2001

[1] Abbreviations: CD, cluster of differentiation; IBV, infectious bronchitis virus; IEL, intraepithelial lymphocyte; IFN-γ, interferon gamma; IL-1, interleukin-1; KLH, keyhole limpet homocyanin; LA, linoleic acid; LNA, linolenic acid; LPS, lipopolysaccharide; NDV, Newcastle disease virus; PUFA, polyunsaturated fatty acid; SRBC, sheep red blood cells.

4. Nutrients regulate pathogen growth and presence

Pathogens require nutrients to proliferate and survive within the host. Pathogens obtain nutrients from the host and minerals are of particular importance to pathogen growth. During periods of infection, the host produces a vast array of acute phase proteins and many of these are metal binding or storage proteins. For example, plasma iron concentrations decrease during infection and this mineral is directed to the liver where it is stored (Laurin *et al.*, 1987). In fact, for many species iron is first limiting for bacterial growth and increased availability of this mineral increases pathogen virulence (Kadis *et al.*, 1984; Knight *et al.*, 1983; Weinberg, 1999). Therefore, it is important to consider the form and amount of mineral nutrition provided to poultry since their repartitioning during periods of infection is an important component to host defense, since iron is important for pathogen growth.

The gut is the largest lymphoid tissue in the body and therefore represents an attractive target for modulating immunity. Gut associated lymphoid tissue can be modulated by feeding supplements of live microorganisms, i.e. probiotics, or complex carbohydrates that stimulate the growth of certain microorganisms, i.e. prebiotics. Feeding probiotics and prebiotics results in improved gut health and increased enteric and systemic immune responses to specific pathogens (Erickson *et al.*, 2000; Rastall *et al.*, 2002). Although the exact mechanisms of how probiotics and prebiotics alter immunity are unclear, their use has been shown to improve humoral, cellular and innate immunity. A better understanding of the mechanisms responsible for their immunomodulatory properties is needed.

Gut microflora populations are responsive to diet composition, feed additives and immune challenges. Microbiota are involved in nutrient synthesis, prevent foreign bacterial colonization and waste degradation. Drastic changes in the gut microbial ecology are detrimental and can result in opportunistic infections. For example, vaccination for Eimeria resulted in a similar microbial population upon challenge, whereas unvaccinated animals challenged with Eimeria had significant changes in the cecal microbial populations (Hume *et al.*, 2006). Essential oil blends have also been shown to modulate microbial growth and prevent microflora changes during immune challenges (Hume *et al.*, 2006).

Previous microflora ecology characterization techniques relied on culture-based models to identify gut organisms. More recent techniques include the amplification of ribosomal DNA followed by direct nucleotide sequencing to identify specific microbial populations. These methods will permit detection of microbes that could not be cultured in laboratory conditions and should serve as a tremendous asset to this field of research. It will be important to begin to understand how probiotics and prebiotics modify the microbiota within the intestine given the cross-talk between microbes and immune cells contained within the intestine (Corthesy *et al.*, 2007).

5. Summary

Nutrition can be used as a tool to promote animal health and welfare. The primary objective of diet formulation is to provide nutrients in the appropriate amounts and proportions that are most ideal for promoting animal performance. The nutrient levels are at the minimum amount to achieve this objective and it is not known if these requirements are suitable for the development, maintenance and use of the immune system. Additionally, some nutrients can regulate the type of immune response and often they are included in the diet at levels above growth requirements. The complexity of pathogens and their different routes of infection make universal recommendations for inclusion of immunomodulatory nutrients in poultry diets difficult. The cross-talk between microbes and the immune system within the digestive tract is important for maintaining gut health and animal performance. Since the digestive tract is the largest immune tissue, it is important to increase our understanding of how diet can modulate the interactions between intestinal microbes and the immune system. The immune system is dynamic and complex, requiring proper care and feeding. Further research examining the impact of nutrition on the immune system is needed, particularly as it relates to the integration of nutrients, diet, feeding regime and genetics.

References

Aledo, J.C. (2004). Glutamine breakdown in rapidly dividing cells: waste or investment? *Bioessays* **26**: 778-785.

Allen, P.C. and Fetterer, R.H. (2002). Interaction of dietary vitamin E with Eimeria maxima infections in chickens. *Poultry Science* **81**: 41-48.

Ardawi, M.S. and Newsholme, E.A. (1983). Glutamine metabolism in lymphocytes of the rat. *Biochemical Journal* **212**: 835-842.

Ardawi, M.S. and Newsholme, E.A. (1985). Metabolism in lymphocytes and its importance in the immune response. *Essays in Biochemistry* **21**: 1-44.

Barnes, D.M., Song, Z., Klasing, K.C. and Bottje, W. (2002). Protein metabolism during an acute phase response in chickens. *Amino Acids* **22**: 15-26.

Chen, C.H., Sowder, J.T., Lahti, J.M., Cihak, J., Losch, U. and Cooper, M.D. (1989). TCR3: A third T-cell receptor in the chicken. *Proceedings of the National Academy of Sciences U.S.A.* **88:** 2351-2355.

Cooper, M.D., Cain, W.A., Van Alten, P.J. and Good, R.A. (1969). Development and function of the immunoglobulin producing system. Effect of bursectomy at different stages of development on germinal centers, plasma cells, immunoglobulins and antibody production. *International Archives of Allergy and Immunology* **35:** 242-252.

Cooper, M.D., Chen, C.L., Bucy, R.P. and Thompson, C.B. (1991). Avian T cell ontogeny. *Advances in Immunology* **50:** 87-117.

Corthesy, B., Gaskins, H. R. and Mercenier, A. (2007). Cross-talk between probiotic bacteria and the host immune system. *Journal of Nutrition* **137 (Suppl 2):** 781S-790S.

Coskun, B., Inal, F., Celik, I., Erganis, O., Tiftik, A. M., Kurtoglu, F., Kuyucuoglu, Y. and Ok, U. (1998). Effects of dietary levels of vitamin A on the egg yield and immune responses of laying hens. *Poultry Science* **77:** 542-546.

Cunningham-Rundles, S. (1998). Analytical methods for evaluation of immune response in nutrient intervention. *Nutrition Reviews* **56 (Pt 2):** S27-37.

Dalloul, R. A., Lillehoj, H. S., Shellem, T. A. and Doerr, J. A. (2002). Effect of vitamin A deficiency on host intestinal immune response to *Eimeria acervulina* in broiler chickens. *Poultry Science* **81:** 1509-1515.

Dalloul, R.A., Lillehoj, H.S., Shellem, T.A. and Doerr, J.A. (2003). Intestinal immunomodulation by vitamin A deficiency and lactobacillus-based probiotic in *Eimeria acervulina*-infected broiler chickens. *Avian Disease* **47:** 1313-20.

De Jonge, W.J., Kwikkers, K.L., Te Velde, A.A., Van Deventer, S.J., Nolte, M.A., Mebius, R.E., Ruijter, J.M., Lamers, M.C. and Lamers, W.H. (2002). Arginine deficiency affects early B cell maturation and lymphoid organ development in transgenic mice. *Journal of Clinical Investigation* **110:** 1539-1548.

Droge, W., Eck, H.P., Gmunder, H. and Mihm, S. (1991). Modulation of lymphocyte functions and immune responses by cysteine and cysteine derivatives. *American Journal of Medicine* **91 (3C):** 140S-144S.

Edinger, A.L. and Thompson, C.B. (2002). Antigen-presenting cells control T cell proliferation by regulating amino acid availability. *Proceedings of the National Academy of Sciences U.S.A.* **99:** 1107-1109.

Erickson, K.L. and Hubbard, N.E. (2000). Probiotic immunomodulation in health and disease. *Journal of Nutrition* **130 (2S Suppl):** 403S-409S.

Fox, C.J., Hammerman, P.S. and Thompson, C.B. (2005). Fuel feeds function: energy metabolism and the T-cell response. *Nature Review Immunology* **5:** 844-852.

Frauwirth, K.A. and Thompson, C. B. (2004). Regulation of T lymphocyte metabolism. *Journal of Immunology* **172:** 4661-4665.

Friedman, A., Bartov, I. and Sklan, D. (1998). Humoral immune response impairment following excess vitamin E nutrition in the chick and turkey. *Poultry Science* **77**: 956-962.

Friedman, A., Meidovsky, A., Leitner, G. and Sklan, D. (1991). Decreased resistance and immune response to *Escherichia coli* infection in chicks with low or high intakes of vitamin A. *Journal of Nutrition* **121**: 395-400.

Fulton, J.E. (2004). Selection for avian immune response: a commercial breeding company challenge. *Poultry Science* **83**: 658-661.

Funk, P.E. and Palmer, J.L. (2003). Dynamic control of B lymphocyte development in the bursa of fabricius. *Archivum Immunologiae et Therapiae Experimentalis (Warszawa)* **51**: 389-398.

Goldsby, R., Kindt, T., Osborne, B. and Kuby, J. (2003). *Immunology*. W.H. Freeman and Company.

Greiner, E.F., Guppy, M. and Brand, K. (1994). Glucose is essential for proliferation and the glycolytic enzyme induction that provokes a transition to glycolytic energy production. *Journal of Biological Chemistry* **269**: 31484-31490.

Halevy, O., Arazi, Y., Melamed, D., Friedman, A. and Sklan, D. (1994). Retinoic acid receptor-alpha gene expression is modulated by dietary vitamin A and by retinoic acid in chicken T lymphocytes. *Journal of Nutrition* **124**: 2139-2146.

Hammond, J. (1944). Physiological factors affecting birth weight. *Proceedings of the Nutrition Society* **2**: 8-12.

Haq, A.U., Bailey, C.A. and Chinnah, A. (1996). Effect of beta-carotene, canthaxanthin, lutein, and vitamin E on neonatal immunity of chicks when supplemented in the broiler breeder diets. *Poultry Science* **75**: 1092-1097.

Hentges, E.J., Marple, D.N., Roland, D.A., Sr. and Pritchett, J.F. (1984). Muscle protein synthesis and growth of two strains of chicks vaccinated for Newcastle disease and infectious bronchitis. *Poultry Science* **63**: 1738-1741.

Huff, G.R., Huff, W.E., Balog, J.M., Rath, N.C. and Izard, R.S. (2004). The effects of water supplementation with vitamin E and sodium salicylate (Uni-Sol) on the resistance of turkeys to *Escherichia coli* respiratory infection. *Avian Disease* **48**: 324-331.

Hume, M.E., Clemente-Hernandez, S. and Oviedo-Rondon, E.O. (2006). Effects of feed additives and mixed *Eimeria* species infection on intestinal microbial ecology of broilers. *Poultry Science* **85**: 2106-2111.

Humphrey, B. and Klasing, K.C. (2005). The acute phase response alters cationic amino acid transporter expression in growing chickens (*Gallus gallus domesticus*). *Comparative Biochemistry and Physiology Part A: Molecular & Integrative Physiology* **142**: 485-494.

Humphrey, B.D. (2005). *Nutrient needs of the immune system*. Mid-Atlantic Nutrition Conference, Timonium, MD.

Humphrey, B.D. and Klasing, K.C. (2004). Modulation of nutrient metabolism and homeostasis by the immune system. *World's Poultry Science Journal* **60:** 90-100.

Humphrey, B.D., Koutsos, E.A. and Klasing, K.C. (2002). Requirements and priorities of the immune system for nutrients. In: *Nutrition biotechnology in the feed and food industries: Proceedings of Alltech's 18th annual symposium.* (Eds. Lyons, T.P. and Jasques, K.A.) Nottingham, UK, Nottingham University Press, pp. 69-77.

Humphrey, B.D., Stephensen, C.B., Calvert, C.C. and Klasing, K.C. (2004). Glucose and cationic amino acid transporter expression in growing chickens (*Gallus gallus domesticus*). *Comparative Biochemistry and Physiology Part A: Molecular & Integrative Physiology* **138:** 515-525.

Humphrey, B.D., Stephensen, C.B., Calvert, C.C. and Klasing, K.C. (2006). Lysine deficiency and feed restriction independently alter cationic amino acid transporter expression in chickens (*Gallus gallus domesticus*). *Comparative Biochemistry and Physiology Part A: Molecular & Integrative Physiology* **143:** 218-227.

Kadis, S., Udeze, F.A., Polanco, J. and Dreesen, D.W. (1984). Relationship of iron administration to susceptibility of newborn pigs to enterotoxic colibacillosis. *American Journal of Veterinary Research* **45:** 255-259.

Kidd, M.T., Peebles, E.D., Whitmarsh, S.K., Yeatman, J.B. and Wideman, R.F., Jr. (2001). Growth and immunity of broiler chicks as affected by dietary arginine. *Poultry Science* **80:** 1535-1542.

Klasing, K.C. and Austic, R.E. (1984). Changes in protein synthesis due to an inflammatory challenge. *Proceedings of the Society of Experimental Biology and Medicine* **176:** 285-291.

Klasing, K.C. and Calvert, C.C. (1999). The care and feeding of an immune system: an analysis of lysine needs. In: *Protein Metabolism and Nutrition.* (Eds. Lobley, G.E., White, A. and MacRae), Wageningen Pers, pp. 253-264.

Knight, C.D., Klasing, K.C. and Forsyth, D.M. (1983). *E. coli* growth in serum of iron dextran-supplement pigs. *Journal of Animal Science* **57:** 387-395.

Korver, D.R. and Klasing, K.C. (1997). Dietary fish oil alters specific and inflammatory immune responses in chicks. *Journal of Nutrition* **127:** 2039-2046.

Kwak, H., Austic, R.E. and Dietert, R.R. (1999). Influence of dietary arginine concentration on lymphoid organ growth in chickens. *Poultry Science* **78:** 1536-1541.

Laurin, D.E. and Klasing, K.C. (1987). Effects of repetitive immunogen injections and fasting versus feeding on iron, zinc, and copper metabolism in chicks. *Biology of Trace Element Research* **14:** 153-165.

Leshchinsky, T.V. and Klasing, K.C. (2001). Divergence of the inflammatory response in two types of chickens. *Developmental and Comparative Immunology* **25:** 629-638.

Leshchinsky, T.V. and Klasing, K.C. (2001). Relationship between the level of dietary vitamin E and the immune response of broiler chickens. *Poultry Science* **80:** 1590-1599.

Leshchinsky, T.V. and Klasing, K.C. (2003). Profile of chicken cytokines induced by lipopolysaccharide is modulated by dietary alpha-tocopheryl acetate. *Poultry Science* **82:** 1266-1273.

Lessard, M., Hutchings, D. and Cave, N.A. (1997). Cell-mediated and humoral immune responses in broiler chickens maintained on diets containing different levels of vitamin A. *Poultry Science* **76:** 1368-1378.

Lin, H., Wang, L.F., Song, J.L., Xie, Y.M. and Yang, Q.M. (2002). Effect of dietary supplemental levels of vitamin A on the egg production and immune responses of heat-stressed laying hens. *Poultry Science* **81:** 458-465.

Malmezat, T., Breuille, D., Capitan, P., Mirand, P.P. and Obled, C. (2000). Glutathione turnover is increased during the acute phase of sepsis in rats. *Journal of Nutrition* **130:** 1239-1246.

Malmezat, T., Breuille, D., Pouyet, C., Mirand, P.P. and Obled, C. (1998). Metabolism of cysteine is modified during the acute phase of sepsis in rats. *Journal of Nutrition* **128:** 97-105.

Muir, W.I., Husband, A.J. and Bryden, W.L. (2002). Dietary supplementation with vitamin E modulates avian intestinal immunity. *British Journal of Nutrition* **87:** 579-585.

Murphy, C. and Newsholme, P. (1998). Importance of glutamine metabolism in murine macrophages and human monocytes to L-arginine biosynthesis and rates of nitrite or urea production. *Clinical Science* **95:** 397-407.

Newsholme, P. (2001). Why is L-glutamine metabolism important to cells of the immune system in health, postinjury, surgery or infection? *Journal of Nutrition* **131 (Suppl):** 2515S-2522S; discussion 252.

Newsholme, P., Costa Rosa, L.F., Newsholme, E.A. and Curi, R. (1996). The importance of fuel metabolism to macrophage function. *Cell Biochemistry and Function* **14:** 1-10.

Newsholme, P., Curi, R., Gordon, S. and Newsholme, E.A. (1986). Metabolism of glucose, glutamine, long-chain fatty acids and ketone bodies by murine macrophages. *Biochemical Journal* **239:** 121-125.

Nicholson, B., Manner, C.K., Kleeman, J. and MacLeod, C.L. (2001). Sustained nitric oxide production in macrophages requires the arginine transporter CAT2. *Journal of Biological Chemistry* **276:** 15881-15885.

Paramithiotis, E. and Ratcliffe, M.J. (1994). B cell emigration directly from the cortex of lymphoid follicles in the bursa of Fabricius. *European Journal of Immunology* **24:** 458-463.

Pink, J.R. (1986). Counting components of the chicken's B cell system. *Immunological Reviews* **91:** 115-128.

Pink, J.R., Vainio, O. and Rijnbeek, A.M. (1985). Clones of B lymphocytes in individual follicles of the bursa of Fabricius. *European Journal of Immunology* **15:** 83-87.

Rastall, R.A. and Maitin, V. (2002). Prebiotics and synbiotics: towards the next generation. *Opinion in Biotechnology* **13**: 490-496.

Rathmell, J.C., Farkash, E. A., Gao, W. and Thompson, C.B. (2001). IL-7 enhances the survival and maintains the size of naive T cells. *Journal of Immunology* **167**: 6869-6876.

Rodriguez, P.C., Zea, A.H., Culotta, K.S., Zabaleta, J., Ochoa, J.B. and Ochoa, A.C. (2002). Regulation of T cell receptor CD3zeta chain expression by L-arginine. *Journal of Biological Chemistry* **277**: 21123-21129.

Rudrappa, S.G. and Humphrey, B.D. (2007). Energy metabolism in developing chicken lymphocytes is altered during the embryonic to posthatch transition. *Journal of Nutrition* **137**: 427-432.

Sijben, J.W., Klasing, K.C., Schrama, J.W., Parmentier, H.K., van der Poel, J.J., Savelkoul, H.F. and Kaiser, P. (2003). Early in vivo cytokine genes expression in chickens after challenge with *Salmonella typhimurium* lipopolysaccharide and modulation by dietary n--3 polyunsaturated fatty acids. *Developmental and Comparative Immunology* **27**: 611-619.

Sijben, J.W., Nieuwland, M.G., Kemp, B., Parmentier, H.K. and Schrama, J.W. (2001). Interactions and antigen dependence of dietary n-3 and n-6 polyunsaturated fatty acids on antibody responsiveness in growing layer hens. *Poultry Science* **80**: 885-893.

Sijben, J.W., Schrama, J.W., Parmentier, H.K., van der Poel, J.J. and Klasing, K.C. (2001). Effects of dietary polyunsaturated fatty acids on in vivo splenic cytokine mRNA expression in layer chicks immunized with *Salmonella typhimurium* lipopolysaccharide. *Poultry Science* **80**: 1164-1170.

Sklan, D., Melamed, D. and Friedman, A. (1995). The effect of varying dietary concentrations of vitamin A on immune response in the turkey. *British Poulty Science* **36**: 385-392.

Suffredini, A.F., Fantuzzi, G., Badolato, R., Oppenheim, J.J. and O'Grady, N.P. (1999). New Insights into the Biology of the Acute Phase Response. *Journal of Clinical Immunology* **19**: 203-214.

Sung, Y.J., Hotchkiss, J.H., Austic, R.E. and Dietert, R.R. (1991). L-arginine-dependent production of a reactive nitrogen intermediate by macrophages of a uricotelic species. *Journal of Leukocyte Biology* **50**: 49-56.

Takahashi, K., Orihashi, M. and Akiba, Y. (1999). Dietary L-arginine level alters plasma nitric oxide and apha-1 acid glycoprotein concentrations, and splenocyte proliferation in male broiler chickens following Escherichia coli lipopolysaccharide injection. *Comparative Biochemistry and Physiology Part C: Pharmacology, Toxicology and Endocrinology* **124**: 309-314.

Tian, S. and Baracos, V.E. (1989). Effect of *Escherichia coli* infection on growth and protein metabolism in broiler chicks (*Gallus domesticus*). *Comparative Biochemistry & Physiology Part A: Molecular & Integrative Physiology* **94**: 323-331.

Tian, S. and Baracos, V.E. (1989). Prostaglandin-dependent muscle wasting during infection in the broiler chick (*Gallus domesticus*) and the laboratory rat (*Rattus norvegicus*). *Biochemistry Journal* **263**: 485-490.

Weinberg, E.D. (1999). Iron Loading and Disease Surveillance. *Emerging Infectious Diseases* **5**: 346-352.

Emerging pathologies: are we prepared?

Steve R. Collett
Poultry Diagnostic and Research Center, University of Georgia, 953 College Station Road, Athens, Georgia 30602, USA

1. Introduction

Many of today's emerging pathologies are not new problems. Although there are some that truly are emerging diseases, most have merely expanded their geographic distribution or re-emerged primarily because of management techniques and production system design constraints.

In an effort to feed a rapidly increasing and informed population cost effectively, producers have had to intensify production by increasing the size and throughput of their systems. These enormous close-confinement rearing systems, designed to maximize productivity by optimizing bird comfort, unfortunately also inherently increase the risk and consequence of disease. The physiological stress of keeping pace with genetic potential for growth lowers disease resistance, making these animals more vulnerable to disease challenge, while the close proximity of susceptible hosts increases the rate of infectious disease spread.

Narrowing profit margins have reduced the producer's tolerance for deviation from expected performance, thus increasing the significance of more subtle or subclinical disease. This pushes producers to improve production and health management practices to consistently achieve expected performance, whilst still satisfying evolving consumer demand for food safety and animal welfare. Unfortunately this has been complicated by consumer pressure to remove many of the health enhancing feed additives that have been instrumental in allowing the intensification process to be successful. Under such stringent constraints previously unimportant diseases have re-emerged as major concerns.

Geneticists have managed to maintain performance improvement, in the straight line response range, over the last few decades. However, each increment in improvement is one step closer to the point of physiological limit and hence, diminishing returns. Unless production systems are modified to satisfy bird requirements, physiological, nutritional and agent induced pathologies will continue to emerge.

2. Important emerging pathologies

Cost efficient production has always been a prerequisite to developing and maintaining a profitable business in the poultry sector. Poultry product prices have been forced down by intense competition for market share. Due to the fact that marketing, retail and wholesale sectors have refused to give up margin, it has been left to the producer to improve efficiency in order to maintain profitability. Economic efficiency has been improved through a combination of consistent annual incremental gain in genetic potential and intensification of the farming operations.

The primary objective of any poultry production system is to optimize the economic efficiency of converting poultry feed into human food. Highly successful breeding and selection programs have provided the platform for annual improvements in the biological efficiency of this feed conversion. However, this potential for very rapid growth in the case of meat type birds and very high and sustained egg production in the case of layers has itself increased the significance of physiological stress, to a point where even minor perturbation in homeostasis has a significant effect on performance. Efforts to convert this potential for biological efficiency into economic efficiency, involve housing these animals in large groups (to limit fixed costs) in highly mechanized units while keeping nutrition as close to requirement as possible (to limit variable costs). It is unfortunately the very success of these improvements in efficiency that has changed the landscape of disease challenge in modern poultry production.

The traditional paradigm of disease has been shaped by the study of specific diseases in individual animals and tends to overemphasize the importance of the infectious agent. In intensive animal agriculture it is the influence of environmental disease determinants that decide the economic outcome of infectious agent challenge. Health and disease are not mutually exclusive and the impact of disease challenge on productivity is apparent long before the clinical signs of disease appear. The focus on flock health management has subtly shifted from avoiding an inadequate immune response that may result in mortality to, avoiding an exaggerated or inappropriate immune response that is costly in terms of productivity. The task has shifted from the diagnosis, treatment and control of specific disease conditions in the individual bird, to preventing and limiting the consequence of more complex multifactorial disease syndromes in order to maximize the productivity of the flock.

Emerging viral diseases present a challenge to the poultry industry because there are no effective treatment options, while emerging bacterial, protozoal and parasitic diseases present a challenge because the treatment options are either no longer available or no longer effective. Clearly the reason for emergence and the approach to controlling diseases within these two categories is very different.

The molecular structure of a virus particle is relatively simple, making immunological recognition very acute and the control of known viral diseases possible through immunization. Provided the immune system has been primed by vaccination, immunological protection against viral disease challenge is usually highly successful. Emerging and re-emerging viral diseases arise when novel or immunologically distinct viruses are introduced into naïve populations (Jones *et al.*, 2008). In the absence of prior exposure, immune recognition and activation is delayed and the extent of the primary immune response is frequently inadequate to prevent clinical disease (Janeway and Travers, 1997). Under such conditions virus replication and spread occurs rapidly with potentially devastating consequences (Capua *et al.*, 2003). While the majority of emerging viral diseases in humans are the result of exposure to novel viruses, it is the emergence of variant strains that pose the biggest threat to the poultry industry (Woolhouse, 2008). Although controlled environment housing and good biosecurity practices have been highly effective in preventing the introduction of novel viruses, increased population density and vaccination are likely to have enhanced the emergence of variant strain viruses. The high population densities provide the opportunity for antigenic shift through gene recombination, whilst vaccination creates positive selection pressure for the variant strains of viruses (Swayne and Halvorson, 2003).

Bacteria and protozoa are, in contrast to viruses, structurally and immunologically complex, making protection through vaccination much less successful. Although a great deal of research effort is and has been focused on developing effective immunization strategies for these diseases, antibiotics have remained the primary means of control (Chapman *et al.*, 2002). This point is well illustrated by the escalating difficulties experienced in the EU with the current systematic withdrawal of in-feed antibiotics (Fabri, 1992). It is no coincidence that the downward trend in prophylactic (in-feed) antibiotic usage has been matched by an increase in therapeutic use (VANTURES, 2005). Many expert committees blame the use of in-feed antibiotics in food animal agriculture for the proliferation of antibiotic resistant strains of bacteria, and for the increase in prevalence of antibiotic resistant infections in humans (JETCAR, 1999).

This is undoubtedly providing the impetus to ban in-feed antibiotic use, even though a link to increased antibiotic resistant bacterial disease in humans has not been conclusively established (Cox and Ricci, 2008). Consumer pressure to remove antibiotics from the food animal nutritionist's arsenal is, however, is likely to continue the trend towards re-emergence of previously controlled bacterial and protozoal diseases. The industry must adapt in order to remain competitive.

Although emerging diseases of catastrophic nature, such as the current H5N1 Highly Pathogenic Avian Influenza (HPAI) outbreak, have attracted the world's attention, it is the more subtle erosive diseases that have accounted for the greatest losses (Collier *et al.*, 2008). The prohibition of in-feed antibiotic growth promoter use in the European Union has initiated the emergence of 'wet litter' problems in broilers across the region. This apparent bacterial enteritis, commonly called dysbacteriosis or sub-clinical Clostridial enteritis, has undermined the economic efficiency of this agricultural operation and made bird welfare management more difficult (Fabri, 1992).

The use of in-feed antibiotic growth promoters has been fundamental in allowing the industry to intensify to the extent that it has. Raising broilers in close confinement without antibiotic growth promoters tips the scales in favour of those gastrointestinal inhabitants that can multiply most rapidly, surviving from one cycle to the next, despite stringent clean-out and disinfection of the house. Bacteria, in particular the Clostridial species, that have a generation interval of hours and the capability to form highly resistant spores, are clear winners in such situations so it is no surprise that clostridial pathologies have re-emerged.

Clostridium perfringens has for a long time been known to cause necrotic enteritis in poultry and has more recently been incriminated but not confirmed to be associated with more subtle intestinal disease and cholangiohepatitis (Ochiai *et al.*, 2003; Wilson *et al.*, 2005). Although the pathologies caused by this organism have been attributed to the alpha toxin, some recent studies have elucidated that it is a newly discovered toxin, aptly named NetB, that is more likely responsible for the destruction of the epithelial lining (Keyburn *et al.*, 2006, 2008). *Clostridium perfringens* is conditioned to produce its repertoire of toxins, except its enterotoxins (cpe), during the exponential growth phase which infers that intestinal disease associated with this organism occurs when conditions in the gut satisfy requirements for rapid multiplication (Petit *et al.*, 1999).

Inflammation of the intestinal epithelium stimulates a self perpetuating cascade of events that allows rapid *Clostridium perfringens* multiplication and toxin production. Inflammation stimulates mucus secretion, and increases paracellular permeability and feed passage (peristalsis) which compromises digestion and facilitates domination of the intestinal flora with mucolytic species such as *Clostridium perfringens* (Deplancke *et al.*, 2002; Collier *et al.*, 2008). Toxin production peaks during this phase of rapid growth and the resulting tissue damage stimulates production of pro-inflammatory cytotoxins which cause further damage to the intestinal lining (Collier *et al.*, 2003). Deteriorating epithelial barrier function allows toxin and antigen penetration, which further increases inflammation (Cooper, 1984; Collier *et al.*, 2003). Once this spiral of events is initiated nothing short of antibiotic therapy will prevent the rapid onset of death as a result of toxemia or septicemia.

The extent of *Clostridium perfringens* induced pathology varies according to how conducive the intestinal conditions are to multiplication and toxin production. *Clostridium perfringens* is thought to be a common inhabitant of the chicken's intestinal tract but its presence is influenced by diet and the intestinal environment (Niilo, 1980; Benno *et al.*, 1988; Deplancke *et al.*, 2002; Lu *et al.*, 2003; Wages and Opengart, 2003; Drew *et al.*, 2004). Clinical necrotic enteritis requires heavy colonization of the small intestine with a high level of toxin production. With a lower level of colonization and toxin production the clinical manifestation of enteric disease is much more subtle. Although the integrity of the intestinal lining is macroscopically normal, its selective permeability and absorptive capacity is compromised. This results in deterioration in feed efficiency and an increase in faecal moisture content. This subclinical enteropathy manifests as suboptimal feed efficiency and wet litter, arguably one of the most important emerging pathologies globally.

'Poultry litter becomes wet when the rate of water addition (urine/faeces/spillage) exceeds the rate of removal (evaporation)' this definition emphasizes that the cause of wet litter is multi factorial in nature (Collett, 2007). Under thermo-neutral conditions, water intake is almost double feed intake and approximately 40% of consumed water is lost via the urine and faeces. Feed intake, water intake and water excretion rate have increased with selection for growth rate. In today's terms at a stocking density of 34 kg/m^2 (~7 lb/ft^2) an average house holds around 20,000 broilers, so at six weeks of age daily water intake is around 6 tonnes. Approximately 2.5 tonnes of this consumed water is deposited into the litter daily and has to be removed by evaporation and ventilation in order to keep litter moisture stable at 25%. With such large volumes of water being deposited into the litter, a relatively minor perturbation in water balance can very rapidly change litter moisture content. Once water excretion rate exceeds the evaporation capacity of the ventilation system, the litter becomes 'wet'. The extent of this problem is proportional to stocking density and is especially evident when the relative humidity is high and or ambient temperatures are extreme.

In small-bird programs where birds are attaining market weight by 34 days, stocking densities reach house-design capacity (~34 kg/m^2) as early as 4 weeks of age. This also happens to be the time when intestinal damage caused by the *Eimeria* spp. of intestinal parasites peaks. Cell damage and the resulting inflammatory response stimulated by these parasites compromises the selective permeability of the intestinal lining resulting in water efflux and increased faecal moisture content. This increase in faecal moisture on its own, is often enough to raise litter moisture to threatening levels. It is however, the effect that this inflammation has on the gut flora in the short term and the house flora in the long term which is most concerning. The initial change in gut flora is often insufficient to affect bird performance in the first grow-out because the 'buffering' capacity of the litter or house flora slows the evolution of gut ecology. However, as the house flora is displaced and replaced by the changing gut flora, the 'buffering' capacity of the house flora diminishes and the compositional change in gut flora becomes sufficient to have a performance effect. So, although the gut flora can change and adapt rapidly, the long term effect is only noticed 3 to 5 grow-outs later, in deep litter systems.

The dynamics of gut microbial communities are difficult to study. Recent advances in molecular techniques have opened eyes to the amount of dialogue that occurs between these gut inhabitants and the gut epithelial cells (Lu *et al.*, 2003, 2006; Kelly *et al.*, 2004). Most of this communication appears to be chemical in nature, the intricacies of which go down to the gene level (Kelly *et al.*, 2004; Ley *et al.*, 2006). This symbiotic relationship has developed over millions of years. Microbial populations are kept in check with quorum sensing molecules, microbial population dynamics are manipulated by host cell secretion, host inflammatory response is controlled by microbial messenger proteins and the very form and function of the intestine is modified by these microbial inhabitants and their metabolites (Collier *et al.*, 2003; Kelly *et al.*, 2004; Waters and Bassler, 2005; Rawls *et al.*, 2006). These intestinal microbial communities traditionally referred to as normal flora, undergo temporal evolution and although fairly host specific, are subject to substrate (ingredient) and intestinal environmental (niche) influence. The pH and redox potential of the upper and lower digestive tract create entirely different ecological niches and consequently support significantly different microbial communities. The very fact that certain components of the normal diet are non-digestible yet essential to a healthy digestive tract, bears testimony to the intricacies of evolutionary design.

The upper digestive tract is designed to function at low pH. Apart from the physiological significance of high hydrogen ion concentration on ingredient digestion and absorption this low pH also provides a hostile environment for potential pathogens. The lower intestinal tract is in contrast designed to function at high pH. These conditions of low hydrogen ion concentration and redox potential favour the microbes capable of fermenting the non-digestible fraction of the diet. Provided digestion and absorption is efficient, competition for nutrients in the distal gut is fierce. It is this shortage of microbial substrate that helps to keep the potentially pathogenic, basophilic members of the distal gut communities in check.

The non-digestible fraction of a natural diet is primarily carbohydrate in nature. If however, digestion and absorption efficiency in the upper tract is compromised, the composition and quantity of microbial nutrients being delivered to the distal intestinal tract changes. The size and composition of the microbial population responds to this substrate change. Initially the increase in nutrient supply supports population growth and the relative increase in protein favours the replication of proteolytic organisms. The community becomes increasingly dominated by these proteolytic organisms and as the metabolite load increases the conditions for organism survival deteriorate. At this point the diversity of the microbial population is reduced and the risk of pathogen challenge to the digestive tract is greatly increased. The competitive exclusion capability of the microbial community is compromised and the environment (pH and redox potential) favours the survival of potential pathogens like *Clostridiun perfringens*.

The risk posed by pathogen colonization of the distal gut is increased by the somewhat unique feature of continual and very efficient reverse peristalsis in the avian species. *C. perfringens*, a toxin producing organism that is capable of very rapid multiplication is well suited to dominating the microbial community under such conditions (Sacranie *et al.*, 2007). Once the competitive advantage swings in favour of this organism the abundance of nutrient substrate allows it to shift rapidly into exponential growth phase and maximize toxin production.

The cascade of events that follows, rapidly results in a life threatening enteropathy (Collier *et al.*, 2003).

Ecological changes in the gut are also reflected in the population dynamics of the more phylogenetically complex intestinal inhabitants such as the protozoa. Some of the motile protozoa like *Cochlosoma anatis*, *Histomonas meleagridis* and *Trichomonas* spp. frequently proliferate in situations where intestinal ecology is destabilized. This is particularly so in turkeys. Although these organisms are thought by some to be primary pathogens of the intestinal tract this is difficult to demonstrate conclusively since they cannot be grown in pure culture (Bermudez, 2003; McDougald, 2003a). Their presence in the gut does, however, indicate compromised gut health and their control has been hampered by the withdrawal of the nitroimidazole drugs from the market (Bermudez, 2003; McDougald, 2003a). Histomoniasis has in particular reemerged as an issue in both turkeys and chickens without chemoprophlaxis and treatment as options (McDougald, 2005). The *Eimeria* spp. that cause coccidiosis are in contrast, primary pathogens of the intestinal tract and coccidiosis remains one of the most costly intestinal diseases of poultry (McDougald, 2003).

Coocidiosis is re-emerging as a primary concern in conventional programs where drug resistance is developing and is of course a major stumbling block in drug free programs (Stephen *et al.*, 1997). The life cycle of these protozoa is short, their replication is very efficient and the oocysts are very resistant, making them well adapted to survive in modern intensive production systems. Even low infection rates can depress bird performance directly, by reducing the efficiency of nutrient assimilation and indirectly by upsetting the ecological balance of the intestinal tract. Innate and acquired immunity remain the primary means of protection but the complex structure of these organisms makes protection, through adaptive immunity, challenging.

3. Implications for poultry production: are we prepared?

Since the goal of a poultry-operation is to convert feed into food as economically as possible, it is critical to manage both the risk and consequence of disease challenges. Whilst the biological efficiency of this *feed conversion* is governed primarily by intrinsic or genetic determinants, under today's intensive production systems it is the extrinsic disease determinants that ultimately decide the efficiency of the operation in both biological and financial terms. Capital investment in the housing's environmental-control capability and the effective operation of these controls are fundamental to economic success.

Even subtle disease challenge such as vaccination with live respiratory disease vaccines can compromise efficiency if exacerbated by environmental disease determinants.

The presence of disease in a flock only becomes significant when the functional derangement of normal metabolic and homeostatic processes causes a decline in productivity, which is sufficiently large to affect economic efficiency. This occurs either as a result of disease induced anorexia or through specific effects on the physiological processes of nutrient metabolism, respiration and excretion. Only in severe cases does disease cause mortality and yet mortality rate is used universally as a measure of flock health status. Health and disease are not mutually exclusive, yet a flock is assumed to be healthy when it is performing to standard and is free of clinical disease. Productivity provides a much more sensitive measure of flock health but since it is the composite output of a population, it gives no indication of the variance in health status between individuals within the flock. Within flock variance provides the most sensitive measure of flock health status but flock uniformity is itself still a fairly coarse measure of health status.

Disease challenge management must be considered as an integral part of any poultry business's risk management program. It involves the development and implementation of a stringent biosecurity plan which comprises a hierarchy of components directed at preventing or limiting the risk and consequence of disease. Economic analysis is a critical step in biosecurity plan design since resource allocation must match risk (Bale *et al.*, 1993). Although it is difficult to accurately determine the precise risk and consequence of a disease challenge it is possible to rank disease challenges according to relative risk (Gifford *et al.*, 1987). From a risk analysis point of view emerging diseases can be considered in two broad categories; those that are unlikely to occur but are catastrophic in nature and those that are highly likely to occur but have less of a financial impact.

Eradication is an appropriate means of risk management for diseases that have a devastating effect on profitability through their impact on productivity or sales (Smith, 1995). Disease control programs are, in contrast, aimed at keeping the prevalence and economic impact of disease at a tolerable level. There is a subtle shift in emphasis from prevention through early detection and elimination, to limiting the consequence or economic impact of the disease, i.e. 'damage control'.

No disease control or eradication program would be successful without diligent surveillance. The objective of the data collection system is to provide incisive and epidemiologically informative indicators which will permit objective judgment and decision making. To support an eradication program, surveillance must be sufficiently intense to detect the source case of an outbreak, so that biocontainment through quarantine and slaughter can be carried out before disease spread occurs. The difficulty lies in confident early detection since this requires frequently testing a large sample of the population. The heavy economic burden of such intense surveillance is difficult to carry and frequently biases risk management decisions, especially when the probability of a disease outbreak is low. Potentially devastating diseases such as HPAI can be effectively eradicated provided adequately robust bioexclusion, surveillance and biocontainment programs are in place. The fact that this disease has public health connotations has helped to justify sufficient commitment to surveillance, the linchpin between bioexclusion and biocontainment.

For disease control purposes a surveillance program is aimed at identifying when disease prevalence changes are sufficient to initiate corrective action. The difficulty is in distinguishing common cause (background variation) from special cause (a disease effect). Surveillance for the purpose of disease control or more appropriately flock health management remains an art. There are no specific tests that can be carried out to determine the health status of a flock, thus placing the emphasis/burden on skilful clinical assessment. Flock health monitoring systems involve a combination of clinical observation and necropsy findings. The sample size and frequency constraints of these procedures severely limit sensitivity, thus emphasising the need for careful sample selection and attention to detail. The focus should be on identifying and eliminating subtle disease challenge since even a mildly exaggerated or inappropriate immune response will compromise performance.

In contrast to respiratory disease, where early signs of disease are outwardly apparent and relatively easy to detect, low grade gastrointestinal disease is much more insidious. Breeding and selection for performance has down regulated the clinical signs of intestinal disease: birds continue to eat and drink at 'normal' levels even when gastrointestinal disease is quite advanced. Early changes in intestinal absorptive capacity, normally indicated by litter moisture changes because of compromised water balance, are often masked by litter buffering capacity and good ventilation. Similarly, accelerated cellular sloughing, usually indicated by the presence of orange mucus in the faeces, is to a degree masked by high feed through-flow rates.

Current health monitoring systems are too focused on cocidiosis lesion scoring and are not sensitive enough to detect early changes in intestinal integrity. Microscopic changes in the intestinal lining can significantly depress feed conversion efficiency but they are difficult to detect even at necropsy. The body works hard to maintain villus height as evidenced by the correlation between crypt depth and epithelial attrition rate. Once villus height starts to decrease, epithelial attrition rate has exceeded regeneration capacity. Intestinal wall thickness and tensile strength are thus fairly coarse measures of intestinal integrity. For early detection of deteriorating gut health the focus needs to shift to subtle changes in the colour and consistency of intestinal content, the amount of mucus and cell debris, the presence or absence of gas and inflammatory exudate. These factors indicate inflammation at the intestinal interface and changes in the intestinal microbiota; which compromise short term and long term performance respectively.

Whilst findings from health monitoring/tracking systems provide the motivation for corrective action it is often the lack of effective intervention options that allows pathologies to emerge. In-feed antibiotics have been used extremely widely to improve feed conversion efficiency (Rosen, 1995). This improvement occurs in part because in-feed medication reduces the number of potentially pathogenic organisms in the intestinal tract, and directly or indirectly down regulates inflammatory response at the intestinal interface. Although low grade intestinal disease is not life threatening, inflammation at the intestinal interface does compromise biological and economic efficiency. By reducing the negative impact of the increased pathogen challenge that occurs with intensification, in-feed medication has been pivotal in allowing intensification to take place. Sudden withdrawal of these in-feed antibiotics allows pathogen colonization of the intestinal tract to proceed unhindered and environmental contamination to increase exponentially with time. Under intensive rearing conditions it is this environmental contamination that becomes a threat to sustained performance. Potentially pathogenic organisms displace and replace the beneficial symbiotic organisms of the gut and with time these changes are reflected in the house flora. With high stocking densities this replacement/displacement occurs rapidly and pathogen numbers soon build to the point where they cause disease. Without the protection of in-feed antibiotics it is the design constraints of the production system that are first limiting and previously controlled diseases such as histomoniasis and coccidiosis once again come to the fore. Vaccination has proven to be a poor substitute for chemoprophylaxis in the case of these diseases, demonstrating that the industry is ill prepared to cope with the withdrawal of antiprotozoal agents from the feed.

Immunologically derived resistance to coccidiosis develops within 4-6 weeks of challenge. The level of natural exposure that typically occurs in modern intensive rearing systems is however, too high. At this level of challenge the tissue damage caused by both the parasite and the ensuing inflammatory response is enough to compromise the digestive process to where it affects performance before immunity develops. Traditional coccidiosis control programmes rely on in feed chemoprophylaxis to suppress the level of natural exposure to a tolerable level (McDougald, 2003b). In the absence of chemoprophylaxis, vaccination is the logical alternative but immunization by vaccination is difficult.

These organisms are relatively large and structurally complex making immune mediated defence complicated and strain specific. In addition, organism replication rate is explosive and attenuation difficult. Because attenuation is difficult, immunization with non-attenuated strains through a process of controlled exposure is common. The efficacy of controlled exposure (non-attenuated strains) and to a lesser extent vaccination (attenuated strains) is highly dose dependant. There is a fine line separating an appropriate from an inappropriate immune response and a relatively small error in vaccine dose can mean the difference between effective immunization and clinical coccidiosis. Overdosing results in excessive tissue damage before the host can mount a protective response, while under-dosing leaves the host unprotected and susceptible to field challenge.

Unfortunately the extent of tissue damage caused by vaccination or controlled exposure is enough to directly (tissue damage) and indirectly (dysbacteriosis) depress performance (growth rate and feed conversion). This is particularly evident and of economic significance in small-bird (<2 kg) programs, where there is insufficient time for compensatory growth. While it is clear that a coccidiosis vaccination program requires support, the design and implementation of effective support strategies are slow in development. Decision makers are ill equipped to evaluate the results of rapidly evolving molecular science and are consequently trapped by their traditional paradigm. This somewhat parochial outlook often results in irrational decision making based on intuition instead of fact, thus hindering progress. Arsenic containing compounds such as 3-nitro-4-hydroxy-phenylarsonic acid (roxarsone) have for example been used in broiler and turkey feed as part of the coccidiosis control program for decades (McDougald *et al.*, 1992) Although the mechanism of action remains elusive these products have 'stood the test of time' and are consequently used with confidence. Newer non-antibiotic ingredients with similar properties are in contrast, often ignored.

Strategies aimed at averting the negative impact of coccidiosis vaccination are no different to the fundamentals of managing gut health. They need to be cost effective, sustainable, farm specific and holistic. Intervention selection needs to be science based but practical and must address the specific objective of reducing coccidiosis induced tissue damage and preventing consequential changes in the composition of the intestinal microbial community.

Modulation of the immune response to cocidiosis challenge is pivotal, so as to enhance protective response and reduce or contain the extent of inflammatory mediated tissue damage. This begins with establishing and maintaining a diverse but stable gut microbial community, as soon after hatch as possible. Microbial diversity is the key because it ensures stability and avoids chaotic blooms of pathogenic organisms when the ecology of the intestine is altered by coccidiosis induced inflammation (Yachi and Loreau, 1999; Collier *et al.*, 2003; Backhed *et al.*, 2005). The composition of intestinal microbiota today, is likely a reflection of centuries of group selection (Ley *et al.*, 2006). The process has been accelerated in the poultry industry by intense breeding and selection programs at the primary breeder level and in-feed antimicrobial use at the broiler level, making changes difficult. Progress in selection for favourable microbial communities is, however, possible over time since group selection pressures on microbial communities are high when they are transferred to a new habitat (Ley *et al.*, 2006). Provided pathogens are kept as minority members of the microbiota they are purged at flock depletion, house clean-out and placement (Travisano and Velicer, 2004). Gut microbial community management becomes a numbers game. Each intervention provides incremental gain and it will frequently take more than one to make meaningful change.

The downstream effects of compromised digestion and absorption can also be countered by improving nutrient availability and/or reducing the nutrient density of the diet. Exogenous enzyme addition to diets with corresponding reduction in nutrient density helps to avert the negative impact of malassimilation on microbiota diversity. There is ample research to demonstrate that the use of exogenous enzymes improves digestibility (Choct and Annison, 1992; Rosen, 2000, 2001). In this respect too little attention has been given to the protein fraction of the diet. This is surprising, since toxic compounds are generated by the proteolytic and ureolytic activity of caecal microbes thus threatening the stability of the microbial community and the integrity of the intestinal lining (Abboud *et al.*, 1997).

To cope without traditional chemoprophylaxis, decision makers need to embrace change and explore the alternatives with an open mind since: 'The problems we have today cannot be solved by thinking the way we thought when we created them' - Albert Einstein.

4. Conclusion

The poultry business is exactly that, a business, the aim of which is to make a profit. Judicious use of cost items like antibiotics and non-antibiotic feed additives makes both scientific and economic sense; and begins with accurate and early diagnosis. The critical steps in addressing any emerging disease are bioexclusion, surveillance and biocontainment. Eradication requires that there is an effective means of detecting infection, containing the infection and preventing dissemination of the disease causing agent (Smith, 1995). The success of any eradication program hinges on early detection through thorough and diligent surveillance and this is possibly where we are not adequately prepared to handle emerging pathologies.

Many of the gastrointestinal diseases are multifactorial in nature and involve subtle shifts in intestinal microbial populations making eradication impractical or impossible. Since there is no specific disease causing agent, eradication through bioexclusion, early detection with classical testing techniques and biocontainment is impossible. With these disease entities the emphasis should be on reducing the consequence of disease. The difficulties are that firstly, diagnosis is based on subjective assessment, requiring experience and attention to detail. Secondly, there is currently insufficient knowledge of the gut ecology to predict how community changes might negatively impact bird performance or how these communities could be positively manipulated. We are thus still poorly prepared to handle emerging pathologies of the gastrointestinal tract.

References

Abboud, P., Gallais, A., Janky, E. and Gidenne, T. (1997). Caeco-colic digestion in the growing rabbit: impact of nutritional factors and related disturbances. *Livestock Production Science* **51**: 73-88.

Backhed, F., Ley, R.E., Sonnenburg, J.L., Peterson, D.A. and Gordon, J.I. (2005). Host-bacterial mutualism in the human intestine. *Science* **307**: 1915-1920.

Bale, M.J., Bennett, M., Berringer, J.E. and Hinton, M. (1993). The survival of bacteria exposed to desiccation on surfaces associated with farm buildings. *Journal of Applied Bacteriology* **75**: 519-528.

Benno, Y., Endo, K. and Mitsuoka, T. (1988). Isolation of fecal *Clostridium perfringens* from broiler chickens and their susceptibility to eight antimicrobial agents for growth promotion. *Nippon Juigaku Zasshi* **50**: 832-834.

Bermudez, A.J. (2003). Cochlosoma anatis Infection. In: Diseases of Poultry. 11th edition (Ed. Saif, Y.M.) Ames, Iowa State Press, USA, pp. 996-1001.

Capua, I., Marangon, S., dalla Pozza, M., Terregino, C. and Cattoli, G. (2003). Avian influenza in Italy 1997-2001. *Avian Diseases* **47**: 839-843.

Chapman, H.D., Cherry, T.E., Danforth, H.D., Richards, G., Shirley, M.W. and Williams, R.B. (2002). Sustainable coccidiosis control in poultry production: the role of live vaccines. *International Journal of Parasitology* **32**: 617-629.

Choct, M. and Annison, G. (1992). The inhibition of nutrient digestion by wheat pentosans. *British Journal of Nutrition* **67**: 123-132.

Collett, S.R. (2007). Strategies to Manage Wet Litter. In: *Proceedings of the 19th Annual Australian Poultry Science Symposium.* Sydney, New South Wales, University Publishing Services, AU, pp. 134-144.

Collier, C.T., Van der Klis, J.D., Deplancke, B., Anderson, D.B. and Gaskins, H.R. (2003). Effects of tylosin on bacterial mucolysis, *Clostridium perfringens* colonization, and intestinal barrier function in a chick model of necrotic enteritis. *Antimicrobial Agents Chemotherapy* **47**: 3311-3317.

Collier, C.T., Hofacre, C.L., Payne, A.M., Anderson, D.B., Kaiser, P., Mackie, R.I. and Gaskins, H.R. (2008). Coccidia-induced mucogenesis promotes the onset of necrotic enteritis by supporting *Clostridium perfringens* growth. *Veterinry Immunology and Immunopathology* **122**: 104-115.

Cooper, B.T. (1984). Small intestinal permeability in clinical practice. *Journal of Clinical Gastroenterology* **6**: 499-501.

Cox, L.A. and Ricci, P.F. (2008). Causal regulations vs. political will: Why human zoonotic infections increase despite precautionary bans on animal antibiotics. *Environment International* **34**: 459-475.

Deplancke, B., Vidal, O., Ganessunker, D., Donovan, S.M., Mackie, R.I. and Gaskins, H.R. (2002). Selective growth of mucolytic bacteria including *Clostridium perfringens* in a neonatal piglet model of total parenteral nutrition. *American Journal of Clinical Nutrition* **76**: 1117-1125.

Drew, M., Syed, N., Goldade, B., Laarveld, B., and Van Kessel, A. (2004). Effects of dietary protein source and level on intestinal populations of *Clostridium perfringens* in broiler chickens. *Poultry Science* **83**: 414-420.

Fabri, T.H.F. (1992). Necrotic enteritis, Clostridial enteritis or dysbacteriosis? In: *Proceedings of XIX World Poultry Congress.* Amsterdam, NL, pp. 580-584.

Gifford, D.H., Shane, S.M., Hugh-Jones, M. and Weigler, B.J. (1987). Evaluation of Biosecurity in Broiler Breeders. *Avian Diseases* **31**: 339-344.

Janeway, C.A. and Travers, P. (1997) *Immunobiology: The Immune System in Health and Disease*, 3rd edition. Garland Publishing Inc., New York, USA.

Joint Expert Technical Advisory Committee on Antibiotic Resistance (1999). The use of antibiotics in food-producing animals: antibiotic-resistant bacteria in animals and humans. Report of the JETCAR, AU (JETACAR). Available at: http://www.health.gov.au/pubs/jetacar.htm.

Jones, K.E., Patel, N.G., Levy, M.A., Storeygard, A., Balk, D., Gittleman, J.L. and Daszak, P. (2008). Global trends in emerging infectious diseases. *Nature* **451:** 990-993.

Kelly, D., Campbell, J.I., King, T.P., Grant, G., Jansson, E.A., Coutts, A.G., Pettersson, S. and Conway, S. (2004). Commensal anaerobic gut bacteria attenuate inflammation by regulating nuclear-cytoplasmic shuttling of PPAR-gamma and RelA. *Nature Immunology* **5:** 104-112.

Keyburn, A.L., Boyce, J.D., Vaz, P., Bannam, T.L., Ford, M.E., Parker, D., Di Rubbo, A., Rood, J.I. and Moore, R.J. (2008). NetB, a new toxin that is associated with avian necrotic enteritis caused by *Clostridium perfringens*. *Public Liberary of Science Pathogens,* **4:** e26.

Keyburn, A.L., Sheedy, S.A., Ford, M.E., Williamson, M.M., Awad, M.M., Rood, J.I. and Moore, R.J. (2006). Alpha-toxin of *Clostridium perfringens* is not an essential virulence factor in necrotic enteritis in chickens. *Infection and Immunity* **74:** 6496-6500.

Ley, R.E., Peterson, D.A. and Gordon, J.I. (2006). Ecological and evolutionary forces shaping microbial diversity in the human intestine. *Cell* **124:** 837-848.

Lu, J., Hofacre, C. and Lee, M. (2006) Emerging technologies in Microbial ecology aid in understanding the effects of monensin in the diets of broilers in regard to the complex disease necrotic enteritis. *Journal of Applied Poultry Research* **15:** 145-153.

Lu, J., Idris, U., Harmon, B., Hofacre, C., Maurer, J.J. and Lee, M.D. (2003). Diversity and succession of the intestinal bacterial community of the maturing broiler chicken. *Applied and Environmental Microbiology* **69:** 6816-6824.

McDougald, L.R. (2003). Coccidiosis. In: *Diseases of Poultry* 11th edition. (Ed. Saif, Y.M.) Ames, Iowa State Press, Iowa, USA, pp. 974-991.

McDougald, L.R. (2003a). Other Protozoan Diseases of the Intestinal Tract. In: *Diseases of Poultry* 11th edition. (Ed. Saif, Y.M.) Ames, Iowa State Press, Iowa, USA, pp. 1001-1009.

McDougald, L.R. (2003b). Protozoal Infections. In: *Diseases of Poultry* 11th edition. (Ed. Saif, Y.M.) Ames, Iowa State Press, Iowa, USA, pp. 973-1023.

McDougald, L.R. (2005). Blackhead disease (histomoniasis) in poultry: a critical review. *Avian Diseases* **49(4):** 462-476.

McDougald, L.R., Gilbert, J.M., Fuller, L., Rotibi, A., Xie, M. and Zhu, G. (1992). How much does roxarsone contribute to coccidiosis control in broilers when used in combination with ionophores? *Journal of Applied Poultry Research* **1:** 172-179.

Niilo, L. (1980). *Clostridium perfringens* in animal disease: a review of current knowledge. *Canadian Veterinary Journal* **21**: 141-148.

Ochiai, K., Handharyani, E. and Umemura, T. (2003). Idiopathic hepatic fibrosis with cholestasis in broiler chickens: immunohistochemistry of hepatic stellate cells. *Avian Pathology* **32**: 425-428.

Petit, L., Gibert, M. and Popoff, M.R. (1999). *Clostridium perfringens*: toxinotype and genotype. *Trends in Microbiology* **7**: 104-110.

Rawls, J.F., Mahowald, M.A., Ley, R.E. and Gordon, J.I. (2006). Reciprocal gut microbiota transplants from zebrafish and mice to germ-free recipients reveal host habitat selection. *Cell* **127**: 423-433.

Rosen, G.D. (1995). Antibacterials in poultry and pig nutrition. In: *Biotechnology in the Animal Feeds and Animal Feeding* (Eds. Wallace, R.J. and Chesson, A.) VCH Verlagsgesellschaft mbH D-69461 Weinheim. **8**: 143-172.

Rosen, G. (2000). Multi-factorial assessment of exogenous enzymes in broiler pronutrition: Target and problems. In: *Proceedings of the 3rd European Symposium on Feed Enzymes*. Noordwijkerhout, NL.

Rosen, G.D. (2001). Multi-factorial efficacy evaluation of alternatives to antimicrobials in pronutrition. *British Poultry Science* **42(S1)**: S104-S105.

Sacranie, A., Iji, P.A., Mikkelsen, L.L. and Choct, M. (2007). Occurrence of reverse peristalsis in broiler chickens. In: *Australian Poultry Science Symposium* (Ed. Pym, R.A.E.). Sydney, New South Wales, University Publishing Service, AU, pp. 161-164.

Smith, R.D. (1995). Cost of Disease. In: *Veterinary Clinical Epidemiology: A Problem Oriented Approach* 2nd edition (Eds. Ranton, B. and Arbor, A.) London, Tokyo, CRC Press, Inc., pp. 227-246.

Stephen, B., Rommel, M., Daugschies, A. and Haberkorn, A. (1997). Studies of resistance to anticoccidials in *Eimeria* field isolates and pure *Eimeria* strains. *Veterinary Parasitology* **69**: 19-29.

Swayne, D.E. and Halvorson, D.A. (2003). Influenza. In: *Diseases of Poultry* 11th edition. (Ed. Saif, Y.M.) Ames, Iowa State Press, Iowa, USA, pp. 135-160.

Travisano, M. and Velicer, G.J. (2004). Strategies of microbial cheater control. *Trends in Microbiology* **12**: 72-78.

VANTURES, The Veterinary Antibiotic Usage and Resistance Surveillance Working Group. (2005). MARAN: Monitoring of Antimicrobial Resistance and Antibiotic Usage in Animals in The Netherlands in 2005. published by VANTURES, NL. Available at: www.cidc-lelystad.nl.

Wages, D.P. and Opengart, K. (2003). Necrotic Enteritis. In: *Diseases of Poultry* 11th edition. (Ed. Saif, Y.M.) Ames, Iowa State Press, Iowa, USA, pp. 781-784.

Waters, C.M. and Bassler, B.L. (2005). Quorum sensing: cell-to-cell communication in bacteria. *Annual Review of Cell and Developmental Biology* **21**: 319-346.

Wilson, J., Tice, G., Brash, M.L. and St. Hilaire, S. (2005). Manifestations of *Clostridium perfringens* and related bacterial enteritides in broiler chickens. *World's Poultry Science Journal* **61**: 435-449.

Woolhouse, M.E. (2008). Epidemiology: emerging diseases go global. *Nature* **451**: 898-889.

Yachi, S. and Loreau, M. (1999). Biodiversity and ecosystem productivity in a fluctuating environment: the insurance hypothesis. *Proceedings of the National Academy of Science, USA* **96**: 1463-1468.

'Zero *Salmonella*': the new European mantra?

*Filip van Immerseel[1], Ulrich Methner[2], Frank Pasmans[1], Freddy Haesebrouck[1]
and Richard Ducatelle[1]*
[1]*Department of Pathology, Bacteriology and Avian Diseases, Research Group
Veterinary Public Health and Zoonoses, Faculty of Veterinary Medicine, Ghent
University, Salisburylaan 133, 9820 Merelbeke, Belgium*
[2]*Friedrich-Loeffler-Institute, Federal Institute for Animal Health, Jena Branch,
Naumburger Str.96a, 07743 Jena, Germany*

1. *Salmonella* as cause of gastrointestinal disease in humans

The number of annual *Salmonella* infections in humans is tremendously high worldwide. A worldwide egg-associated salmonellosis pandemic started in the '70s and is currently fading away, thanks to huge efforts by policy makers and the poultry industry. This pandemic has been caused by the serotype *Salmonella enteritidis*. Due to its preferential association with laying hen eggs, combined with the way humans tend to store (room temperature), handle and eat (non-cooked) eggs, *Salmonella enteritidis* had and still has a major impact on human health. In the European Union (EU) 165,023 cases of salmonellosis were reported in 2006, which represents an incidence of 34.6 per 100,000 persons (EFSA, 2007a). In 2006, there was a 7.6% decrease in incidence from 2005, and this was part of a significant decreasing trend over the past three years. Of the human infections caused by *Salmonella* in 2006, *Salmonella enteritidis* was identified in about 60% of the cases to be the cause of the infection, and *Salmonella typhimurium* in 14% of the cases. Other serotypes responsible human illness cause for each serotype, less than 2% of the human infections. Other serotypes in the top 10 of causes of human salmonellosis cases in the EU are; Infantis, Virchow, Newport, Hadar, Stanley, Derby, Agona and Kentucky. While the serotype *typhimurium* causes human contamination due to consumption of pig and to a lesser extent, poultry meat, serotypes such as Infantis, Hadar and Virchow are typically associated with broiler meat. While total EU *Salmonella* contamination levels have decreased the last few years, the antimicrobial resistance of the *Salmonella* isolates is still increasing. *Typhimurium* serotypes in particular is causing concern, as about 40% of all *Salmonella typhimurium* strains isolated in 2006 were resistant to 4 or more antimicrobials. Only ciprofloxacin resistance was still very low (0.7%). It is clear that differences exist in contamination levels, serotype distributions and trends of *Salmonella* contamination levels between the different EU countries.

2. Layers and *Salmonella*: *Enteritidis* is the predominant serotype to be controlled

Although, but to a lesser extent, other serotypes can also infect and colonize laying hens, Enteritidis is the predominant serotype found in eggs. Several EU member states have reported data from investigations of table eggs and the overall EU prevalence in 2006 was 0.8% (EFSA, 2007a). More than 90% of all egg-isolates were strains of the serotype Enteritidis. The other 10% of the isolates were strains from different serotypes but mostly isolated in only one EU member state, indicating that the importance of non-*Salmonella enteritidis* isolates in eggs is marginal. The high prevalence of serotype Enteritidis in table eggs is in not completely consistent with the serotype distribution in laying hens. In 2006, 4.8% of the EU laying hen flocks were found to be *Salmonella* positive and about 75% of all isolates were serotype Enteritidis strains (EFSA, 2007a). More than 10% were *Salmonella typhimurium* strains, and a range of other serotypes was found in laying hen flocks. A large-scale baseline study of the European Food Safety Authority (EFSA) in 2005 revealed the presence of *Salmonella* spp. in 30.7% of 4561 large-scale laying hen holdings in the EU (EFSA, 2006a). The actual percentage of positive laying hen flocks in the same year reported by the member states revealed a percentage of *Salmonella* positive flocks in the EU of 3.2%. Thus, the prevalence of 4.8% in 2006 is most likely a serious underestimation, the sampling methods and analytical methods used to analyze the samples were clearly more sensitive in the baseline study. More than 51% of all *Salmonella* isolates in the EFSA baseline study were *Salmonella enteritidis* strains. The fact that different non-Enteritidis serotypes can be isolated from 25-50% of the *Salmonella* infected laying hen flocks, while more than 90% of all isolates from eggs are serotype Enteritidis strains (the other 10% are derived from a minority of member states), indicates that the serotype Enteritidis harbors some intrinsic characteristics that lead to a specific interaction with either the reproductive tract of chickens, or the egg components. The exact mechanisms of these important traits are currently unclear, but recent data, which is outlined below, reveals some important differences between Enteritidis and other serotypes.

Generally, eggs can be contaminated by *Salmonella* on the outer shell and inside the egg. The former could potentially occur due to the presence of *Salmonella* in the hen's environment or passage of the egg through the cloaca. The latter could be caused as a consequence of either shell penetration or colonization of the reproductive tract of laying hens and thus incorporation in the forming egg.

2.1. Outer shell contamination

After oviposition, any contaminated environment in the area of the laid egg can result in outer shell contamination. It is clear that any bacterial species can contaminate the outer shell and that all *Salmonella* serotypes can be present on the outer shell surface, provided that they have contaminated the environment of the layer flock or hatchery. Braden *et al.* (2006) states that intensive control measures in the US, examining eggs for cracks and washing and disinfecting eggs, have not eliminated egg contamination with *Salmonella enteritidis*, indicating that internal egg contamination is a major issue with respect to contamination of humans.

2.2. Internal egg contamination

2.2.1 Eggshell and membrane penetration

In order to contaminate the egg content *Salmonella* bacteria contaminating the outer eggshells have to pass the cuticle, which is a hydrophobic proteinaceous layer, the crystalline eggshell that contains pores and the shell membranes. The presence of *Salmonella* on the shell of eggs, combined with cooling of moist, freshly laid eggs from the body temperature of the hen to the ambient air temperature is believed to cause bacterial passage through the shell and even the inner shell membranes. It is clear that eggshell and membrane penetration are not a unique property of *Salmonella enteritidis*, other *Salmonella* serotypes and totally unrelated bacteria can traverse these barriers in the used models (De Reu *et al.*, 2006). This means that either eggshell penetration does not take place frequently in practice, or that *Salmonella enteritidis*, of all strains that can penetrate the eggshell, is the only one that can survive and multiply once inside the albumen.

2.2.2 Reproductive tract colonization

In many reports (Timoney *et al.*, 1989; Shivaprasad *et al.*, 1990; Humphrey *et al.*, 1991; Thiagarajan *et al.*, 1994; Keller *et al.*, 1995) contamination of eggs is associated with isolation of *S. enteritidis* from the oviduct, suggesting a contamination of the egg in this location. Oviduct colonization can be the result of ascending infections from the cloaca, descending infections from the ovary, and/or systemic spread of *Salmonella*. In natural infections, the albumen has been shown to be the principal site of contamination, but due to the low numbers of contaminated eggs after natural infections and the fact that it is labor-intensive to study this topic, not enough studies have been performed to accurately confirm this statement (Humphrey *et al.*, 1991). In experimental infection trials, most authors report the albumen but others report the yolk, as most frequently contaminated. Although eggs can most likely be contaminated by penetration of the eggshell, it is clear that *Salmonella* also can contaminate eggs after colonization of the ovary or oviduct and thus become incorporated in the forming egg. Depending on the site of reproductive tract colonization, *Salmonella* would be incorporated in the egg at different locations. Infection of the ovary will result in yolk contamination, while contamination of the albumen is believed to occur during passage of the egg through the oviduct. Yolk contamination however, is believed not to lead to the production of contaminated eggs, as the yolk is very nutrient-rich and *Salmonella* bacteria would evoke yolk degeneration while multiplying in this environment. The oviduct contains different segments having different functions and *Salmonella* bacteria infecting the oviduct could be incorporated in either the albumen or the eggshell membranes, depending on the site of colonization (magnum or isthmus, respectively). *Salmonella* bacteria have been found on the mucosal surface and within epithelial cells (of isthmus and magnum), lining the oviduct in naturally and experimentally infected hens (Hoop and Pospischil, 1993; De Buck *et al.*, 2004). It is therefore possible that the bacteria are persistently carried within these cells and at certain time points contaminate forming eggs when released from these cells into the oviduct lumen, as a result of host triggers such as stress or immune suppression. In a recent study, it was shown that after an intravenous infection of laying hens, the ability of 2 serotype Enteritidis strains to colonize the reproductive organs was significantly higher compared with strains of serotypes Heidelberg, Virchow and Hadar strains but not with Typhimurium (Gantois *et al.*, in press). It is possible that *Salmonella enteritidis* has a selective advantage over many other serotypes for oviduct colonization.

2.2.3 Survival in the forming egg

Salmonella bacteria that contaminate the forming eggs in the reproductive tract of laying hens can be incorporated in the egg, provided that presence of the bacteria does not lead to degeneration of the egg contents and provided that the bacteria are not killed by the albumen components. Interestingly, there are differences in the contamination rate of forming eggs and of eggs post-lay. Contamination of the forming eggs during passage through the oviduct is not correlated with the production of *Salmonella* positive eggs. The discrepancy between the isolation rate of *Salmonella* between forming and laid eggs is clearly illustrated by Keller *et al.* (1995), showing that after oral inoculation of a high dose of *S. enteritidis*, about 30% of the forming eggs were positive, in contrast with only 0.6% of freshly laid eggs. The ovum spends about 26 hours in the oviduct, of which 21 hours is in the uterus, where the shell is deposited. It is proposed that the antibacterial properties of the albumen control *Salmonella* contamination at this site, where the temperature is 41 °C. Incorporation of *Salmonella* in the albumen could lead to the bacteria being killed and thus the production of non-contaminated eggs. Indeed, the bacteria encounter multiple antimicrobial components, including lysozyme and ovotransferrin and a high pH. Egg albumen components most likely induce cell wall and DNA damage in *Salmonella*. Survival in the forming egg could be a reason for selective isolation of the serotype *Enteritidis* in laid eggs, if this serotype would have increased expression of genes related to egg albumen resistance. After inoculation of laying hens with 3 different *Enteritidis* and *Typhimurium* strains, it was shown that both serotypes were equally effective in contaminating the forming eggs but only *Enteritidis* was found in the eggs after oviposition (Keller *et al.*, 1997). Clavijo *et al.* (2006) showed that survival in egg albumen at 37 °C was higher for the serotype *Enteritidis* compared with *Typhimurium* and *Escherichia coli*, but not enough isolates were compared to draw conclusions in this study.

2.2.4 Migration from the albumen to the yolk post-lay

There is strong evidence that *Salmonella* bacteria that are introduced in the egg albumen are able to migrate to and penetrate the vitteline membrane, in order to reach the yolk. There they would gain access to a pool of amino acids and other molecules that are necessary for its survival and growth. This would be the major reason for the delay in growth of contaminated eggs until a few weeks of storage at 20 °C. Once *Salmonella* is present in the yolk, the bacteria can grow exponentially. Interestingly, this does not lead to changes in colour, smell or consistency of the egg contents (Humphrey *et al.*, 1993). Different virulence factors, such as flagella and curli fimbriae are thought to play a role (Cogan *et al.*, 2004).

Although the exact mechanism for the specific association of the *Salmonella* serotype *Enteritidis* with the oviduct and/or eggs is unknown, it is clear that one should focus on combating this serotype in layers to prevent egg contamination.

3. Broilers and *Salmonella*: a variety of serotypes, which ones are important?

In broiler flocks, 3.4% of all sampled were found to be *Salmonella* positive in the EU in 2006, which is a slight decrease compared to 2005 (4.1%) and 2004 (4.9%) (EFSA, 2007a). This data was provided by the member states themselves. A *Salmonella* baseline survey, carried out under supervision of EFSA from October 2005 to September 2006, observed a mean EU *Salmonella* prevalence of 23.7% (EFSA, 2007b). This strong discrepancy between the data from the baseline survey and the data reported by the member states is most likely caused by more sensitive sampling and more sensitive analytical methods used. This thus strongly points to a serious underestimation of the actual *Salmonella* prevalence as reported by the individual member states. The five most frequently isolated *Salmonella* serotypes at EU level (in the baseline study) were *Enteritidis*, (37.1%), *Infantis* (20.4%), *Mbandaka* (7.9%), *Typhimurium* (4.6%) and *Hadar* (4.1%). Many other different serotypes are circulating in the broiler population as compared with laying hen flocks. Data indicates that about 5.6% of all EU broiler meat samples analyzed in 2006 were *Salmonella* positive. When the data from Hungary are excluded (more than 50% of all EU broiler meat samples were from Hungary), *Salmonella enteritidis* was the most frequent meat-contaminating serotype, followed by: *Paratyphi B var. Java*, *Infantis*, *Bredeney* and *Typhimurium*. Besides these, many other serotypes can contaminate broiler meat. With respect to prevention of human *Salmonella* infections, in theory all serotypes should be controlled in primary poultry production, as all of these can potentially be transmitted to humans by meat contamination in the slaughterhouse. There is however not a clear relationship between the serotype distribution in broiler flocks or broiler meat and the proportion of human *Salmonella* infections that are caused by consumption of broiler meat (relative to egg consumption), cannot be easily determined. On the other hand, serotypes typically found in broiler flocks and meat but not in other animal species (such as serotypes *Hadar*, *Infantis* and *Virchow*) cause a low proportion of human salmonellosis cases. Other serotypes often found in broiler meat, in contrast, are not frequently causing human salmonellosis. This makes it difficult to speculate about the importance of *Salmonella* strains and serotypes present in broiler flocks with regard to human contamination. In any case, prevention of gut colonization is the main focus in the primary production, as discussed further. All *Salmonella* serotypes can invade intestinal epithelial cells due to the presence of the so-called *Salmonella* Pathogenicity Island I (SPI-1) on the chromosome. SPI-1 expression is needed for gut colonization, which explains why all serotypes colonize the gut in broilers and can thus contaminate meat.

4. European legislation: obligating monitoring schemes and control plans to reach defined targets

As a consequence of the high number of human salmonellosis cases in the EU caused by consumption of chicken products, in 1992, the Council Directive 92/117/EC was adopted. This directive concerned measures for protection against specified zoonoses and specified zoonotic agents in animals and products of animal origin in order to prevent outbreaks of food-borne infections. The directive laid down rules for the collection of information on zoonoses and zoonotic agents, including the relevant measures to be taken in the Member States and at the Community Level. The directive mainly focused on detection. Furthermore, the Council Directive 92/117/EC only laid down rules for the monitoring of *Salmonella* in breeding flocks and not for layer or broiler flocks kept for egg or meat production. The whole set of rules defines sampling schemes, sampling spots, number of samples and reporting of results. Also measures to be taken in cases of infected flocks were specified. It was stated that, when the presence of *Salmonella enteritidis* or *Salmonella typhimurium* is confirmed in the birds of a house, no bird is allowed to leave the house concerned unless the competent authority has authorized the slaughter and destruction under supervision or slaughter in a slaughterhouse designated by the competent authority. Moreover, it was stated that non-incubated eggs produced by the birds in the house in question must be destroyed on the spot or after appropriate marking, be taken under supervision to an approved egg-processing establishment to be heat treated. All birds in the house must be slaughtered and the official veterinarian of the slaughterhouse informed of the decision to slaughter. Where eggs for hatching from flocks in which the presence of *Salmonella enteritidis* or *Salmonella typhimurium* has been confirmed are still present in a hatchery, they must be destroyed or treated as high risk material.

In the early 90s it became clear that the measures taken at the level of breeding flocks, aiming to reduce vertical transmission of *Salmonella*, were insufficient to decrease the contamination level in layers and broilers. As a consequence, no effects were seen on the number of human salmonellosis cases. This clearly indicated that horizontal transmission within layer and broiler flocks is very important for the colonization of the birds by *Salmonella*. It also became clear that epidemiological data on zoonoses and zoonotic agents submitted by the member states were incomplete and difficult to compare between different countries.

As a consequence, in 2003 the EU issued a Directive (2003/99/EC) on the monitoring of zoonoses and zoonotic agents and a Regulation (2160/2003/EC) on the control of *Salmonella* and other specified food-borne zoonotic agents. The purpose of Directive 2003/99/EC was to ensure that zoonoses, zoonotic agents and related antimicrobial resistance are properly monitored and that food-borne outbreaks received proper epidemiological investigation, to enable the collection in the Community of the information necessary to evaluate relevant trends and sources. Member States were obliged to assess trends and sources of zoonoses, zoonotic agents and antimicrobial resistance in their territory. Each Member State needs to transmit to the Commission every year by the end of May a report on trends and sources of zoonoses, zoonotic agents and antimicrobial resistance. Detailed rules concerning the assessment of those reports, including the formats and the minimum information that they must include are given in detail in Directive 2003/99/EC. The Commission then sends the reports to the European Food Safety Authority (EFSA), who examines them and publishes, by the end of November, a summary report on the trends and sources of zoonoses, zoonotic agents and antimicrobial resistance in the community. Member States were also obliged to designate national reference laboratories for each field (where a Community reference laboratory has been established) and inform the Commission thereof. This Directive has led to a sudden increase in monitoring and detection of *Salmonella* and other zoonotic agents. The collected information is published in the 'Community Reports on Trends and Sources of Zoonoses, Zoonotic Agents and Antimicrobial Resistance in the European Union'.

In 2003, the European Parliament and the Council of the European Union also introduced Regulation (EC) No 2160/2003 to ensure that proper and effective measures are taken to detect and control *Salmonella* and other zoonotic agents at all relevant stages of production. Including processing and distribution, particularly at the level of primary production - in feed, in order to reduce their prevalence and the risk they pose to public health. The regulation states that member states should set up national control programs to ensure that targets are reached, within set deadlines. The EU should then make a decision on whether to approve the proposed national control programmes. National control programmes should provide for the detection of zoonoses and zoonotic agents in accordance with certain minimal sampling requirements. Sampling and detection of all *Salmonella* serotypes with public health significance should be done according to the following schemes: (a) breeding flocks: day-old, 4 weeks of age, 2 weeks before transport to the laying unit and every 2 weeks during lay; (b) laying hens: day-old, 2 weeks before transport to the laying unit and every 15 weeks during lay; (c) broilers: before transport to the slaughterhouse. Certain specific measures should also be implemented, these include; that in case of an infection with *Salmonella enteritidis* or *typhimurium* in breeding flocks, non-incubated hatching eggs should be destroyed or used for human consumption only following treatment, in a manner that guarantees the elimination of *Salmonella enteritidis* and *typhimurium*. All birds from these flocks must be slaughtered or destroyed, even the day-old chicks. Eggs derived from these birds that are still present in a hatchery also have to be destroyed or treated as described above. Another specific requirement that should be covered by national control programmes is that from December 2009 onwards, eggs must not be used for human consumption as fresh table eggs unless they originate from a commercial layer flock subject to a national control programme. Moreover, eggs originating from flocks with unknown health status, suspected of being infected or from infected flocks may only be used for human consumption if treated in a manner that guarantees the elimination of all *Salmonella* serotypes with public health significance. Concerning broilers, national control programmes should ensure that fresh poultry meat is not placed on the market for human consumption unless *Salmonella* is absent in 25 grams, starting from December 2010. National control programmes should also allow for the progress made under their provisions to be evaluated and should cover the feed production, the primary production and the processing as well as the preparation of food.

The Commission consulted the EFSA on the use of antimicrobials and vaccines for the control of *Salmonella* in poultry as part of national control programs. As a result of that consultation Commission Regulations 1177/2006 (and 1091/2005) were issued. Regulation 1177/2006 states that live *Salmonella* vaccines shall not be used in the framework of national control programmes where the manufacturer does not provide an appropriate method to distinguish bacteriologically wild-type strains of *Salmonella* from vaccine strains. Live *Salmonella* vaccines shall not be used in the framework of national control programmes in laying hens during production unless the safety of the use has been demonstrated. Vaccination programmes against *Salmonella enteritidis* to reduce the shedding and contamination of eggs, shall be applied, at least during rearing, to all laying hens at the latest from 1 January 2008. This applies to all Member States as long as they did not demonstrated a prevalence below 10 % based on the results of the baseline study. Regulation 1177/2006 also states that antimicrobials shall not be used as a specific method to control *Salmonella* in poultry. The provisions referred to in this Regulation do not apply to substances, micro-organisms or preparations authorized for use as feed additives in accordance Regulation (EC) No 1831/2003 on additives for use in animal nutrition.

In order to set the community targets for *Salmonella* the EU member states have to reach, as stated in Regulation 2160/2003/EC, comparable data on the prevalence of *Salmonella* in the populations of both laying hens and broilers in the member states should be available. Such information was not available and there was a need for special studies to be carried out in order to monitor the prevalence of *Salmonella* in laying hens and broilers. This should be carried out over an appropriate period of time in order to take into account possible seasonal variations. These decisions and details on the setup of these studies are defined in Commission Decisions 2004/665/EC and 2005/636/EC for layers and broilers respectively. In these commission decisions, it is stated that the EU shall undertake technical studies aiming to estimate prevalence of *Salmonella spp.* across the European Union, in flocks of laying hens for table egg production (at the end of the production period) and broilers. The results would be used to set Community targets as provided for in Regulation (EC) No. 2160/2003. Details on the sampling frame, the collection and reporting of data and the EU financial contribution are also specified in these commission decisions.

The Community targets that have to be reached for breeding flocks, laying hen flocks and broiler flocks are specified in different Commission Regulations (Commission Regulations 1003/2005, 1168/2006 and 646/2007). For broilers and layers, baseline studies have been carried out by the Member States (coordinated by EFSA), in order to estimate the prevalence of *Salmonella* spp. (across the EU) in the respective flocks. These results were used to set Community targets. The reductions aimed at by setting community targets are important in view of the strict measures which are to apply to infected laying hen and broiler flocks in accordance with Regulation (EC) No 2160/2003 from December 2009 and 2010 onwards (details see above). Regulation (EC) No 2160/2003 states that *the Community target is to include a numerical expression of the maximum percentage of epidemiological units remaining positive and/or the minimum percentage of reduction in the number of epidemiological units remaining positive*. The maximum time limit within which the target must be achieved and the definition of the testing schemes necessary to verify achievement of the target, should be included. It is also to contain a definition, where relevant, of serotypes with public health significance.

Commission Regulation 1003/2005 implements Regulation (EC) No 2160/2003 regarding Community targets for the reduction of the prevalence of certain *Salmonella* serotypes in breeding flocks of *Gallus gallus*. The Community target for the reduction of *Salmonella enteritidis, Salmonella hadar, Salmonella infantis, Salmonella typhimurium* and *Salmonella virchow* in breeding flocks of *Gallus gallus* should be a reduction of the maximum percentage of adult breeding flocks, comprising at least 250 birds, remaining positive to 1% or less by 31 December 2009. Monitoring and sampling schemes are also defined in Regulation 1003/2005.

Commission Regulation 1168/2006 implements Regulation (EC) No 2160/2003 as regards a Community target for the reduction of the prevalence of certain *Salmonella* serotypes in laying hens of *Gallus gallus*. Regulation 1168/2006 states that the Community target referred to in Regulation (EC) No 2160/2003 for the reduction of *Salmonella enteritidis* and *Salmonella typhimurium* in adult laying hens of *Gallus gallus* shall be as follows: (a) an annual minimum percentage of reduction of positive flocks of adult laying hens equal to at least: (1) 10% if the prevalence in the preceding year was less than 10%; (2) 20% if the prevalence in the preceding year was between 10 and 19%; (3) 30% if the prevalence in the preceding year as between 20 and 39%; (4) 40% if the prevalence in the preceding year was 40% or more: or; (b) a reduction of the maximum percentage to 2% or less. The first target should be achieved in 2008 based on the monitoring starting in the beginning of that year. The actual targets for the Member states can thus be calculated based on the results of the baseline study of EFSA. Regulation 1168/2006 also defines sampling schemes and protocols and detection methods to be used.

Commission Regulation 646/2007 implements Regulation (EC) No 2160/2003 as regards a Community target for the reduction of the prevalence of certain *Salmonella enteritidis* and *typhimurium* in broilers. The Community target, as referred to in Regulation (EC) No 2160/2003, for the reduction of *Salmonella enteritidis* and *Salmonella typhimurium* in broilers shall be a reduction of the maximum percentage of flocks of broilers remaining positive, for these serotypes, to 1% or less by 31 December 2011. Where the presence of *Salmonella enteritidis* and *Salmonella typhimurium* is not detected but antimicrobials or bacterial growth inhibitory effects are detected, it shall be considered as an infected flock of broilers. Regulation 646/2007 also defines sampling schemes and protocols and detection methods to be used.

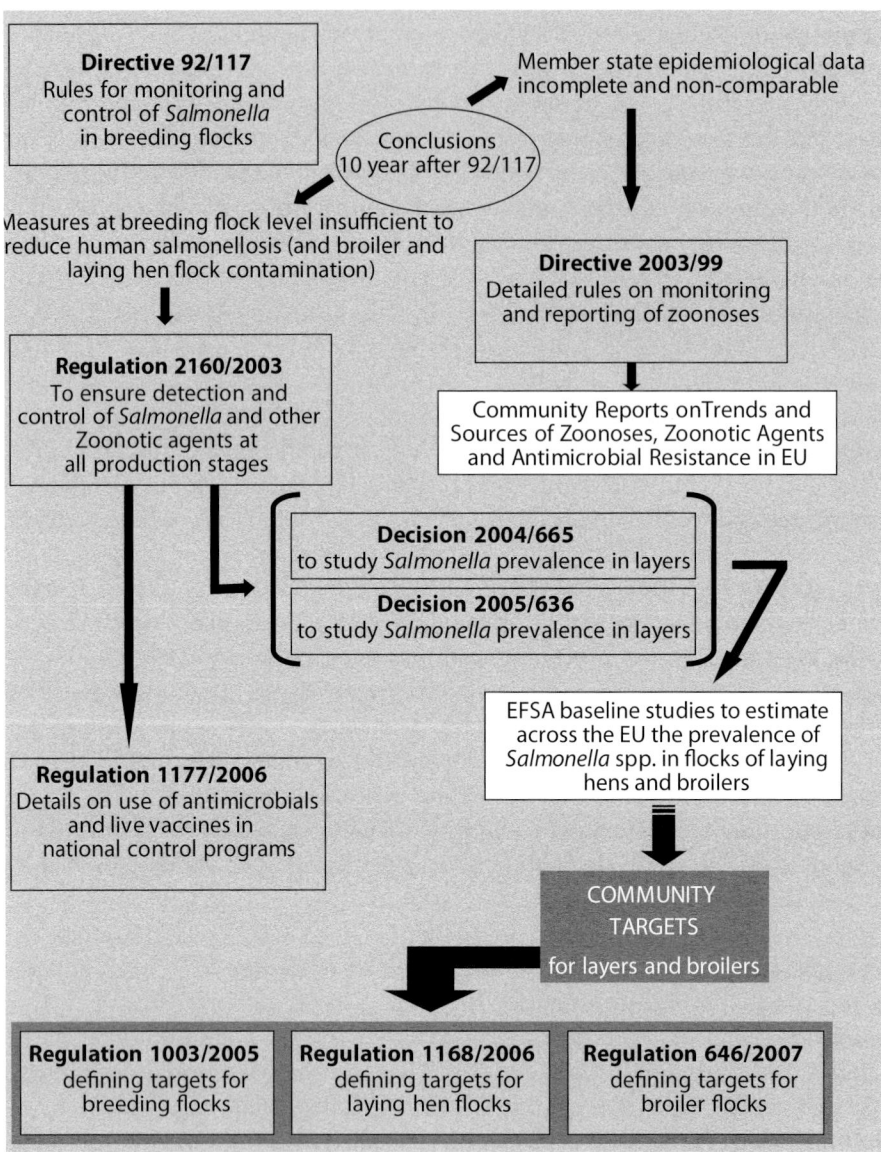

Figure 1. Overview of European Directives, Decisions and Regulations, related to monitoring and control of Salmonella *in chickens.*

5. Vaccination: an efficient way to control *Salmonella enteritidis* in layers

Vaccination has been implemented in national control programmes in some EU member states as a consequence of the implementation of the EU legislations, as discussed above. *Vaccination can be done using live or inactivated vaccines, they should (a) reduce or prevent the intestinal colonisation resulting in reduced faecal shedding and thus egg shell contamination and (b) prevent systemic infection resulting in a decreased colonization of the reproductive tissues, in this way reducing internal egg contamination.* Currently used commercial vaccines in the EU mainly claim to reduce shedding. It is very well documented that both killed and live vaccines can reduce shedding of *Salmonella* in poultry and has been reviewed by Van Immerseel *et al.* (2005).

5.1. Salmonella *vaccines*

5.1.1 Inactivated vaccines

Only one commercial inactivated *S. enteritidis* based vaccine against *S. enteritidis* infection in breeders and laying type chickens is used in EU countries and one commercial inactivated bivalent *S. enteritidis* and *S. typhimurium* dual vaccine has been authorised (Clifton-Hadley *et al.*, 2002). These killed vaccine types are based on bacterial cells cultured under conditions of iron depletion. The commercially available vaccine containing inactivated *S. enteritidis* and *S. typhimurium* (Salenvac T) decreased shedding in a seeder-bird challenge model, when the vaccine was given intramuscularly at day 1 and week 4 of age (Clifton-Hadley *et al.*, 2002). Less than 30% of the vaccinated birds shed *Salmonella* at 10 days post-challenge, in contrast to 80% of the unvaccinated birds. Research groups have also shown that egg contamination caused by systemic spread of the bacteria in the host can be decreased by vaccination using inactivated strains. It was shown that intramuscular vaccination with the inactivated vaccine Salenvac T at day 1, week 4 and week 18 decreased egg contamination after intravenous challenge with *S. enteritidis* (Woodward *et al.*, 2002). Inactivated vaccines are often used in parent flocks. Parenteral administration of inactivated *Salmonella* vaccines to breeding birds will induce a strong production of antibodies and these antibodies will be transferred to the progeny. The maternally transferred antibodies persist for a few weeks in the chicks and although there seems to be some protective effect against disease in the early post-hatch period, there is little effect on intestinal colonisation by challenge strains (Methner *et al.*, 1994; Methner and Steinbach, 1997). It is therefore possible to immunise effectively day-old chicks from vaccinated breeder birds with live *Salmonella* vaccines.

5.1.2 Live vaccines

Although a number of different live *Salmonella* strains have been tested for their efficacy in experimental or semi-field studies only a few are authorised and commercially available for use in poultry in Europe. The accessible live *S. typhimurium* and *S. enteritidis* vaccine strains are either auxotrophic double-marker mutants derived through chemical mutagenesis (Meyer *et al.*, 1993; Springer *et al.*, 2000) or developed on the basis of the principle of metabolic drift mutations (Linde *et al.*, 1997; Hahn, 2000). Some of these live vaccines have been further characterised by molecular methods (Schwarz and Liebisch, 1994). Another live vaccine authorised for prophylactic use against *S. enteritidis* is based on a rough strain of *S. gallinarum* 9R without further molecular characterisation (Feberwee *et al.*, 2001). This vaccine is registered for prophylactic use against *S. enteritidis* and has been shown to reduce the flock level incidence of *S. enteritidis* infections in a large scale study in The Netherlands (Feberwee *et al.*, 2001). While the incidence was 2.5% (2/80 flocks) in vaccinated flocks, the incidence in unvaccinated flocks was 11.5% (214/1854 flocks). More recently, Gantois *et al.* (2006) showed that oral vaccination with the live vaccines TAD *Salmonella* vac®E, TAD *Salmonella* vac®T and a combination of both live vaccine strains, at day 1, week 4 and week 16, decreased internal organ colonization, including reproductive tract colonization and egg contamination. Although it is very difficult to prove reduction of egg contamination following vaccination under field conditions owing to the low and variable percentage of contaminated eggs laid. In various countries a significant decrease in *Salmonella* prevalence in laying hen flocks and human salmonellosis cases, followed implementation of vaccination programmes in layers.

5.1.3 Autologous vaccines

Although most of the *Salmonella* vaccines used in Europe are of commercial origin, autologous vaccines can be applied in some countries. An autologous vaccine is made by isolating a local strain of *Salmonella* spp. from a poultry house or animals and a specific inactivated vaccine is produced for this poultry farm. These vaccines comprise extracts of a culture killed by various processes (heat, formalin, etc.) and various adjuvants. Their efficacy is very controversial and these types of vaccines can only be produced for the farm of concern.

5.2. Improvement of Salmonella *vaccines*

The *Salmonella* vaccines currently registered for use in the EU member states have been authorised on the basis of national regulations, the mutual recognition procedure and several international guidelines and therefore, fulfil the relevant requirements. However, in order to improve the efficacy of *Salmonella* vaccination in poultry, an ideal *Salmonella* vaccine (strain) should possess following characteristics:

- a high degree of protection against systemic and intestinal infection;
- against a variety of important serovars (serogroups);
- adequate attenuation for poultry, other animal species, humans and the environment as well as animal welfare issues;
- the inactivated and live vaccines should not affect the growth of the animal;
- vaccine strains should not be resistant to antibiotics;
- vaccines should be easy to be administer and need to have markers facilitating the differentiation from *Salmonella* wild-type strains;
- application of vaccines should not interfere with *Salmonella* detection methods;
- humoral antibody response after vaccination should be distinguishably from a *Salmonella* wild-type response to allow the use of serological detection methods.

It is clear that *Salmonella* vaccines are very useful in laying hen flocks and can contribute to a decrease in colonization, shedding and egg contamination, when the above mentioned vaccine characteristics are fulfilled. Recently multiple scientific groups have reported a phenomenon, in which oral administration of *Salmonella* wild type and attenuated strains can confer resistance to infection by a virulent *Salmonella* challenge strain within 24 h of administration. This 'competitive exclusion'-like phenomenon is called colonization-inhibition (Nógrady *et al.*, 2003; Van Immerseel *et al.*, 2005; Bohez *et al.*, 2007). These data suggest that it might be possible to administer live *Salmonella* vaccine strains to newly hatched chicks, such that they would colonize the gut extensively and very rapidly. This would induce a profound resistance to colonization by other *Salmonella* strains of epidemiological significance, which may be present in the poultry house or may also have arisen from the hatchery (Van Immerseel *et al.*, 2005). Colonization of the gut by the colonization-inhibition strains (live vaccines strains) would prevent gut colonization by virulent strains, while invasion in the gut tissue would evoke an inflammatory response that would prevent invasion to the internal organs by virulent strains. Therefore, these two characteristics can be included in the list of vaccine criteria:

- attenuated live *Salmonella* vaccine strains should be able to induce the colonization inhibition effect between *Salmonella* organisms;
- attenuated *Salmonella* vaccine strains should have preserved the ability to invade the gut.

In view of the complete list of both already realized and novel characteristics the key question in developing the ideal live *Salmonella* vaccine strain is to find the balance between an accepted level of attenuation and an unaffected ability to induce protection. This will be possible only by using the ability of molecular genetics to produce defined deletion mutants.

6. Other control methods in the primary production and their relevance

A combination of different preventive and curative measures is necessary for the control of *Salmonella* infections in broilers during the primary production phase. First of all, good farming and hygienic practices need to be implemented, in order to avoid introduction of *Salmonella* on the farm or to reduce infection pressure when *Salmonella* is present. Hygienic measures at all levels of the production chain; pre-harvest (during life), harvest (catching and transport) and post-harvest, is essential for successful *Salmonella* control. Hygienic measures should take into account; feed, birds, management, cleaning and disinfection. This can include physical and chemical decontamination treatments of feed, drinking water and the environment of the birds, etc. Eradication of contaminated flocks has not been shown to be an effective measure and the high density of broiler farms within the EU is a major handicap. In the EU decontaminating treatments of chicken meat or eggs, are prohibited. Therefore, prevention and monitoring during the life phase is very important in Europe and additional control measures to increase the resistance of the birds against *Salmonella*, to decrease the shedding and colonization by *Salmonella* field strains are of the utmost importance. Vaccination of the parent broiler flocks can be used to decrease the susceptibility of the offspring by stimulating an immune response through maternal antibodies (Hassan *et al.*, 1996; Methner *et al.*, 1997). Also the use of a genetically more resistant chicken line can help to control *Salmonella* (Kramer *et al.*, 2001; Sadeyen *et al.*, 2004). More importantly however, will be the application of control products in laying hen or broiler flocks. Feed additives are of importance in this regard, as they can have effects on intestinal colonization by *Salmonella*. They are mainly of importance in broiler flocks, as these are the only measures that can be applied in these animals during the live phase, as vaccination is not practical due to the low slaughter age and thus the lack of time to build up a protective immune response: although the colonization-inhibition principle can in theory be used. There are an impressive number of commercially available compounds that can be used as feed or drinking water additives to control *Salmonella*. Some have well documented effects, others are less well documented. Most have been selected by trial and error and the observed effects are on an empirical basis. Since *Salmonella* is a bacterial infection, the most obvious tool for the control of these infections is to use antibiotics. The prophylactic or curative use of antibiotics for the control *Salmonella* infections in poultry however is prohibited under the EU Regulation 2160/2003, amended by EU Regulation 1177/2006. Therefore, most known products applied to control *Salmonella* in broilers flocks are acidic compounds, short- and medium-chain fatty acids, prebiotics and probiotics (used less in the field).

Prebiotics are non-digestible feed ingredients that beneficially affect the host by selectively stimulating the growth and/or activity of one or a limited number of bacterial species already resident in the colon and thus attempt to improve host health (Gibson and Roberfroid, 1995). In this definition it is understood that certain bacterial species multiplying in the colon can have a beneficial effect on host health. These bacterial species can also be administered through the feed and are referred to as probiotics (see below). Most prebiotics are carbohydrates, they can be mono-, di-, oligo- or polysaccharides; either natural or synthetic.

A number of monosaccharides including glucose and fructose are digestible and thus not considered prebiotics according to the standard definition. Mannose is by far the most commonly used monosaccharide feed additive (Allen *et al.,* 1997). Mannose has the saccharide ligand for bacterial type 1 (sef21) fimbriae, which are common surface projections of *Salmonella*, which the bacteria use to attach to the intestinal mucosa. Mannose binds type 1 fimbriae of *Salmonella* and thereby blocks adhesion of type 1 fimbriae bearing bacteria to epithelial cells and mucus (Craven *et al.,* 1992; Dibb-Fuller *et al.,* 1999). It has been shown in early studies that *Salmonella typhimurium* exerts a D-mannose-sensitive cytotoxic effect on the mucosal epithelium of isolated intestinal segments of day-old chicks (Droleskey *et al.,* 1994). Supplementing 2.5% mannose in the feed has been shown to reduce *Salmonella* colonization (Oyofo *et al.,* 1989). Addition to the feed of the more economically relevant 0.1% dose, also reduced shedding, caecal and liver colonization after infection of 2 week old chickens with 2×10^7 cfu of a *Salmonella enteritidis* strain (Agunos *et al.,* 2007).

Amongst many available disaccharides, lactose has been experimentally used in multiple studies. Several studies report that lactose administration increases the short-chain fatty acids and lactic acid concentrations in the gut (Hinton *et al.,* 1990; Hume *et al.,* 1992). Most studies use a combined administration of lactose and competitive exclusion cultures (Hinton *et al.,* 1991; Corrier *et al.,* 1993; Nisbet *et al.,* 1993, 1994). Lactose has been shown to be mainly successful in *Salmonella* reduction during feed withdrawal periods, i.e. during forced moult in layers or during feed withdrawal period before slaughter in broilers. Reductions in *Salmonella* colonization in animals given lactose as drinking water additive are however, in all studies marginal or even non-existing (Barnhart *et al.,* 1999).

Oligosaccharides usually are obtained through enzymatic synthesis or hydrolysis of polysaccharides. Fructo-oligosaccharides (FOS) are short-chain polymers of beta 1-2 linked fructose units, often produced by hydrolysis of inulin. Their use in broiler feed was shown to reduce colonization of the intestine by *Salmonella* in some studies, but mostly the effects were marginal or absent (Bailey *et al.,* 1991; Waldroup *et al.,* 1993; Oyarzabal *et al.,* 1996; Chambers *et al.,* 1997; Fukata *et al.,* 1999). Batch cultures inoculated with faecal slurries show bifidogenic effects of FOS (Rossi *et al.,* 2005), as most bifidobacteria can grow on FOS (Rossi *et al.,* 2005). The effects of bifidobacteria on *Salmonella* colonization are indirect and are explained below. The most probable mechanism of the proposed anti-*Salmonella* activity of FOS is therefore due to an alteration in the microbiota composition of the gut and the production of metabolites, such as short-chain fatty acids. Galacto-oligosaccharides (GOS) are Glu α 1-4 [β Gal 1-6]$_n$, which are not broken down in the stomach or small intestine. GOS also have been shown to selectively stimulate bifidobacteria in rats and pigs (Holma *et al.,* 2002; Tzortzis *et al.,* 2005) but studies in poultry using GOS are absent. Mannan-oligosaccharides (MOS) are derived from the yeast cell wall fragments of *Saccharomyces cerevisiae* are commercially available for use in poultry feed. Mixing 4000 ppm of Bio-Mos® (Alltech Inc, USA) in the feed reduced caecal *Salmonella typhimurium* concentrations after experimental infection of broiler chicks (Spring *et al.,* 2000). Caecal contents from hens fed Bio-Mos, protected chicks from colonization with *Salmonella enteritidis* (Fernandez *et al.,* 2000). The mechanism of action may be through blockage of type 1 fimbriae mediated *Salmonella* adhesion to the mucosa. Modulation of the local (mucosal) immune system and preservation of intestinal wall integrity is also important. Bio-Mos supplementation induces alterations in numbers of lactobacilli, enterococci or anaerobes in the caeca (Spring *et al.,* 2000). In a recent study by Baurhoo *et al.* (2007), Bio-Mos inclusion in feed for chickens increased the numbers of lactobacilli and decreased the *Echerichia coli* load in the gut. More recently a wide range of other oligosaccharides, including isomalto-oligosaccharides, soy-oligosaccharides and xylo-oligosaccharides have been investigated for their prebiotic and health protecting effects in laboratory animal models and human volunteers (reviewed by Tuohy *et al.*, 2005). Little is known however, on their potential for protection against *Salmonella* in poultry.

Another common polysaccharide prebiotic used in chicken feed probably is guar gum (from the seeds of *Cyamopsis tetragonolobus*) and partially hydrolyzed guar gum (PHGG). Feed containing 250 ppm PHGG protected chicks against *Salmonella enteritidis* from 14 days onwards. This is thought to be due to improvements in the balance of the intestinal microbiota (Ishihara *et al.*, 2000). Inulin is a polymer of fructose moieties with a heterogeneous degree of polymerization ranging from 3 to 60 and can be fermented to butyrate (Rossi *et al.*, 2005). It is believed to act in a similar way as FOS but its use as feed supplement in chickens is limited.

Taken together, most prebiotics with a proven protective activity against *Salmonella* act through a shift in the composition of the intestinal microbiota, or through a modification of the metabolic activity of some indigenous intestinal micro organisms. An alternative approach thus can be to directly administer these so-called beneficial micro-organisms. Probiotics by definition are live microbial feed supplements which beneficially affect the host animal by improving its intestinal microbial balance (Fuller, 1989). The original idea was launched over 30 years ago, when Nurmi and Rantala (1973) administered a suspension of gut contents derived from healthy adult chickens to newly hatched chicks, thereby protecting the chicks against gut colonization by *Salmonella enteritidis*. This concept is called competitive exclusion, commercial products based on this concept have been on the market for a number of years and are very effective (Nakamura *et al.*, 2002; Ferreira *et al.*, 2003). One major handicap of this concept however is that the microbiota, which are included in these products, are undefined. Therefore, these competitive exclusion products are not commonly included in feed. Over the years considerable efforts have been made to identify specific micro-organisms that confer a similar protection. There are many reports on the successful application of lactobacilli as probiotics, beneficial not only to growth and performance but also to the resistance of broilers against *Salmonella* infections (Jin *et al.*, 1996; Mulder *et al.*, 1997; Pascual *et al.*, 1999; Van Coillie *et al.*, 2006). This suggests a profound effect of these microorganisms on the intestinal ecosystem, however a limited number of probiotic products are available on the market.

Acidic compounds are used more and more to combat *Salmonella* infections, not only for drinking water acidification. They can also acid release in the proximal gastro-intestinal tract, when added in powder form as a feed additive, or in the distal parts of the gastro-intestinal tract, when coated or encapsulated acids are used in feed. Medium chain fatty acids (MCFA) are strongly bactericidal towards many gram-positive and gram-negative bacteria, including *Salmonella* (Nakai and Siebert, 2003). Even at concentrations as low as 10 mM MCFA still show a bacteriostatic effect on *Salmonella* (Van Immerseel *et al.,* 2004). Short chain fatty acids (SCFA) are the major bacterial fermentation products in the large intestine (Tuohy *et al.,* 2005) and are also commonly added to feed and drinking water. At high concentrations (1%) these products have an antimicrobial effect in moist environment. This microbial growth inhibition is traditionally explained by the ability of these acids to pass across the bacterial cell membrane in undissociated form, dissociate in the more alkaline bacterial cell interior and thereby acidify the bacterial cell cytoplasm (Kashket, 1987). When the acid-treated feed is eaten by the chickens, it is both warmed and moistened and thus the activity of the SCFA should increase. It has long been known however that up to 95% of the SCFA produced during carbohydrate fermentation may be taken up and utilized by the host (Cummings *et al.,* 1987). Thus SCFA added to feed in powder form or drinking water may exert their activity in the lumen of the crop and gizzard but not further down the gastrointestinal tract (Thompson and Hinton, 1997) however, coated and encapsulated products, are on the market. These formulations aim to bring the SCFA further down in the gastro-intestinal tract and to release the acids in the area commonly colonized by *Salmonella*. For drinking water acidification and powder form feed additives, formic and acetic acid are most widely used. In coated and encapsulated products butyrate is of particular importance, as it provides part of the daily energy requirements of the gastrointestinal mucosa, playing an important role in proliferation and differentiation of the epithelial cells (Macfarlane *et al.,* 1997). Butyrate upregulates the expression of tight junction proteins, thereby enhancing the barrier function of the intestinal epithelium (Bordin *et al.,* 2004). It also inhibits inflammation through the NFκB pathway (Hodin, 2000). Sodium butyrate administered in feed in concentrations up to 0.2% increases feed conversion ratio, daily weight gain and intestinal villus length in broilers (Hu and Guo, 2007). Butyrate and propionate (as opposed to acetate and formate), at concentrations similar to those naturally produced in the lower gastrointestinal tract of the chicken, reduce invasiveness of *Salmonella* in intestinal epithelial cells in vitro (Van Immerseel *et al.,* 2004). This phenomenon of reduced invasion by butyrate is mediated by a specific down-regulation of the genes encoded on the *Salmonella* pathogenicity island

1 (Gantois *et al.,* 2006). Coating butyric acid on a carrier protects the acid from absorption in the upper digestive tract, transporting the acid down to the caeca, where *Salmonella* bacteria are known to colonize and invade the mucosa. As expected, the use of coated butyric acid in feed, protects chickens from caecal colonization of *Salmonella* (Van Immerseel *et al.,* 2005). Considering the above mentioned characteristics, it may be advantageous to enhance butyric acid production by the endogenous microbiota, using prebiotics.

7. Can we reduce *Salmonella* prevalence in chicken flocks to zero?

Due the recent EU Regulations and Directives that were published by the EU and the implementation of the regulations by the member states, the awareness of the governments and the poultry industry to control *Salmonella* in poultry has increased. Since the strict Community Targets and deadlines at which the targets need to be reached have been set, member states have established or are establishing control plans for primary production. It is clear that the number of layer and broiler flocks contaminated with *Salmonella* will most likely decrease in the future due to the established action plans. As an example, after introduction of the obligatory vaccination programme for layers in Belgium, the prevalence in layer flocks decreased and the number of human *Salmonella enteritidis* cases decreased by more than 70% over 3 years. Although not all member states of the EU report this kind of spectacular decreases, the overall trend is a decrease in numbers of contaminated layer flocks. In broilers, some countries have huge contamination levels and a decreasing trend has not yet been observed. It is a utopia to eradicate *Salmonella* completely, bringing the prevalence of *Salmonella* in broiler and layer flocks to zero, as the pathogen is a commonly found bacterium in the environment. *Salmonella* will therefore, most likely never be eradicated in the poultry industry but the main issue is to keep the flock prevalence, the within-flock prevalence and the numbers of bacteria in infected animals at such a low level, that contamination of eggs and meat, and thus transmission to humans, is a rare event. This will require the coordinated action of governments and poultry-related industries; including the feed industry and slaughterhouses. Therefore, the control plans will need to encompass biosecurity and hygienic measures at different stages of the production chain, combined with the use of control measures. Concerning layer flocks, vaccination is a very good tool to control *Salmonella* to a certain level. For broilers, vaccination is not an option and the use of effective feed additives is an important tool.

It is clear that this will be an enormous effort that will cost a huge amount of money. Furthermore, we will need to be constantly aware of the probability that new serotypes will emerge. In addition, a possibly low prevalence in the future and consequent drop in the effort to control *Salmonella* will definitely result in a re-emergence. Indeed, *Salmonella* bacteria can be carried within a host at low levels, although being prevented to multiply in the gut and consequently shedding at a high level, by the applied control measures. A large-scale study carried out by the authors showed that despite the fact that overshoes and faecal samples were *Salmonella* negative in most flocks, analysis of caecal samples from the same flocks, in about 50% of the cases, were positive for *Salmonella*, in 1 to 5% of the animals. Moreover, even zero prevalence in the EU monitoring schemes will not guarantee that *Salmonella* is not present within a poultry flock. The *Salmonella* bacterium, once the EU can keep it under control to a reasonable level, will wait for prevention and control measures to be weakened and strike back.

References

Agunos, A., Ibuki, M., Yokomizo, F. and Mine, Y. (2007). Effect of dietary beta1-4 mannobiose in the prevention of *Salmonella* Enteritidis infection in broilers. *British Poultry Science* **48**: 331-241.

Allen, V.M., Fernandez, F. and Hinton, M.H. (1997). Evaluation of the influence of supplementing the diet with mannose or palm kernel meal on *Salmonella* colonization in poultry. *British Poultry Science* **38**: 485-488.

Anonymous (1992). Council Directive 92/117/EEC of 17 December 1992 concerning measures for protection against specified zoonoses and specified zoonotic agents in animals and products of animal origin in order to prevent outbreaks of food-borne infections and intoxications

Anonymous (2003a). Directive 2003/99/EC of the European Parliament and of the Council of 17 November 2003 on the monitoring of zoonoses and zoonotic agents, amending Council Decision 90/424/EEC and repealing Council Directive 92/117/EEC.

Anonymous (2003b). Regulation (EC) No 1831/2003 of the European Parliament and of the Council of 22 September 2003 on additives for use in animal nutrition.

Anonymous (2003c). Regulation (EC) No 2160/2003 of the European Parliament and of the Council of 17 November 2003 on the control of *Salmonella* and other specified food-borne zoonotic agents.

Anonymous (2004). Commission Decision of 22 September 2004 concerning a baseline study on the prevalence of *Salmonella* in laying flocks of *Gallus gallus*.

Anonymous (2005a). Commission Decision of 1 September 2005 concerning a financial contribution by the Community towards a baseline survey on the prevalence of *Salmonella* spp. in broiler flocks of *Gallus gallus* to be carried out in the Member States.

Anonymous (2005b). Commission Regulation (EC) No 1003/2005 of 30 June 2005 implementing Regulation (EC) No 2160/2003 as regards a Community target for the reduction of the prevalence of certain *Salmonella* serotypes in breeding flocks of *Gallus gallus* and amending Regulation (EC) No 2160/2003.

Anonymous (2005c). Commission Regulation (EC) No 1091/2005 of 12 July 2005 implementing Regulation (EC) No 2160/2003 of the European Parliament and of the Council as regards requirements for the use of specific control methods in the framework of the national programmes for the control of *Salmonella*.

Anonymous (2006a). Commission Regulation (EC) No 1168/2006 of 31 July 2006 implementing Regulation (EC) No 2160/2003 as regards a Community target for the reduction of the prevalence of certain *Salmonella* serotypes in laying hens of *Gallus gallus* and amending Regulation (EC) No 1003/2005.

Anonymous (2006b). Commission Regulation (EC) No 1177/2006 of 1 August 2006 implementing Regulation (EC) No 2160/2003 of the European Parliament and of the Council as regards requirements for the use of specific control methods in the framework of the national programmes for the control of *Salmonella* in poultry.

Anonymous (2007). Commission Regulation (EC) No 646/2007 of 12 June 2007 implementing Regulation (EC) No 2160/2003 of the European Parliament and of the Council as regards a Community target for the reduction of the prevalence of *Salmonella enteritidis* and *Salmonella typhimurium* in broilers and repealing Regulation (EC) No 1091/2005.

Bailey, J., Blankenship, L.C. and Cox, N.A. (1991). Effect of fructooligosaccharide on *Salmonella* colonization of the chicken intestine. *Poultry Science* **70**: 2433-2438.

Barnhart, E.T., Caldwell, D.J., Crouch, M.C., Byrd, J.A., Corrier, D.E. and Hargis, B.M. (1999). Effect of lactose administration in drinking water prior to and during feed withdrawal on *Salmonella* recovery from broiler crops and ceca. *Poultry Science* **78**: 211-214.

Bohez, L., Dewulf, J., Ducatelle, R., Pasmans, F., Haesebrouck, F. and Van Immerseel, F. (2008). The effect of oral administration of a homologous hilA mutant strain on the long-term colonization and transmission of *Salmonella* Enteritidis in broiler chickens. *Vaccine* **26**: 372-378.

Bordin, M., D'Atri, F., Guillemot, L. and Citi, S. (2004). Histone deacetylase inhibitors up-regulate the expression of tight junction proteins. *Molecular Cancer Research* **2**: 692-701.

Braden, C.R. (2006). *Salmonella* enterica serotype Enteritidis and eggs: a national epidemic in the United States. *Clinical Infectious Diseases* **43**: 512-517.

Chambers, J.R., Spencer, J.L. and Modler, H.W. (1997). The influence of complex carbohydrates on *Salmonella* Typhimurium colonization, pH, and density of broiler ceca. *Poultry Science* **76**: 445-451.

Clavijo, R.I., Loui, C., Andersen, G.L., Riley, L.W. and Lu, S. (2006). Identification of genes associated with survival of *Salmonella* enterica serovar Enteritidis in chicken egg albumen. *Applied and Environmental Microbiology* **72**: 1055-1064.

Clifton-Hadley, FA, Breslin, M, Venables, LM, Springs, KA, Cooles, SW, Houghton, S, Woodward, MJ. (2002). A laboratory study of an inactivated bivalent iron restricted *Salmonella enterica* serovars Enteritidis and Typhimurium dual vaccine against Typhimurium challenge in chickens. *Veterinary Microbiology* **89**: 167-179.

Collard, J-M., Bertrand, S., Dierick, K., Godard, C., Wildemauwe, C., Vermeersch, K., Duculot, J., Van Immerseel, F., Pasmans, F., Imberechts, H. and Quinet, C. (2008). Drastic decrease of *Salmonella* Enteritidis isolated from humans in Belgium in 2005, shift in phage types and influence on foodborne outbreaks. *Epidemiology and Infection* **6**: 771-781.

Corrier, D.E., Hargis, B.M., Hinton, A.J. and DeLoach, J.R. (1993). Protective effects of used poultry litter and lactose in the feed ration on *Salmonella* Enteritidis colonization of leghorn chicks and hens. *Avian Diseases* **37**: 47-52.

Craven S.E., Cox, N.A., Bailey, J.S., Blankenship, L.C. (1992). Binding of *Salmonella* strains to immobilized intestinal mucosal preparations from broiler chickens. *Avian Diseases* **36**: 296-303.

Cummings, J.H., Pomare, E.W., Branch, W.J., Naylor, C.P.E. and Macfarlane, G.T. (1987). Short chain fatty acids in human large intestine, portal, hepatic and venous blood. *Gut* **28**: 1221-1227.

De Buck., J., Pasmans, F., Van Immerseel, F., Haesebrouck, F. and Ducatelle, R. (2004b). Tubular glands of the isthmus are the predominant colonization site of *Salmonella* Enteritidis in the upper oviduct of laying hens. *Poultry Science* **83**: 352-358.

De Reu, K., Grijspeerdt, K., Messens, W., Heyndrickx, M., Uyttendaele, M., Debevere, J. and Herman, L. (2006). Eggshell factors influencing eggshell penetration and whole egg contamination by different bacteria, including *Salmonella* Enteritidis. *International Journal of Food Microbiology* **112**: 253-260.

Dibb-Fuller, M.P., Allen-VErcoe, E., Thorns, C.J., Woodward, M.J. (1999). Fimbriae- and flagella-mediated association with and invasion of cultures epithelial cells by *Salmonella* Enteritidis. *Microbiology* **145**: 1023-1031.

Droleskey, R.E., Oyofo, B.A., Hargis, B.M., Corrier, D.E. and DeLoach, J.R. (1994). Effect of mannose on *Salmonella* Typhimurium-mediated loss of mucosal epithelial integrity in cultures chick intestinal segments. *Avian Diseases* **38**: 275-281.

EFSA, European Food Safety Authority. (2006). Preliminary report on the analysis of the baseline study on the prevalence of *Salmonella* in laying hen flocks of *Gallus gallus*. *The EFSA Journal* **81**: 1-71.

EFSA, European Food Safety Authority. (2007a). The Community Summary Report on Trends and Sources of Zoonoses, Zoonotic agents, Antimicrobial Resistance and Foodborne Outbreaks in the European Union in 2006.

EFSA, European Food Safety Authority. (2007b). Report of the Task Force on Zoonoses Data Collection on the Analysis of the baseline survey on the prevalence of *Salmonella* in broiler flocks of *Gallus gallus*, in the EU, 2005-2006. Part A: *Salmonella* prevalence estimates. *The EFSA Journal* **98**: 1-85.

Feberwee, A., De Vries, T.S., Hartman, E.G., De Wit, J.J., Elbers, A.R.W. and De Jong, W.A. (2001). Vaccination against *Salmonella* Enteritidis in Dutch commercial layer flocks with a vaccine based on a live *Salmonella* Gallinarum 9R strain: evaluation of efficacy, safety, and performance of serologic *Salmonella* tests. *Avian Diseases* **45**: 83-91.

Fernandez, F., Hinton, M. and Van Gils, B. (2000). Evaluation of the effect of mannan-oligosaccharides on the competitive exclusion of *Salmonella* Enteritidis colonization in broiler chicks. *Avian Pathology* **29**: 575-581.

Fukata, T., Sasai, K., Miyamoto, T. and Baba, E. (1999). Inhibitory effects of competitive exclusion and fructooligosaccharide, singly and in combination, on *Salmonella* colonization of chicks. *Journal of Food Protection* **62**: 229-233.

Fuller, R. (1989). Probiotics in man and animals. *Journal of Applied Bacteriology* **66**: 365-378.

Gantois, I. Ducatelle, R., Pasmans, F., Hautefort, I., Thompson, A., Hinton, J.C. and Van Immerseel, F. (2006a). Butyrate specifically down-regulates *Salmonella* pathogenicity island 1 gene expression. *Applied and Environmental Microbiology* **72**: 946-949.

Gantois, I., Ducatelle, R., Timbermont, L., Bohez, L., Haesebrouck, F., Pasmans, F. and Van Immerseel, F. (2006b). Oral immunisation of laying hens with the live vaccine strains of TAD *Salmonella* vac E and TAD *Salmonella* vac T reduces internal egg contamination with *Salmonella* Enteritidis. *Vaccine* **24**: 6250-6255.

Gibson, G and Roberfroid, M. (1995). Dietary modulation of the human colonic microbiota: introducing the concept of prebiotics. *Journal of Nutrition* **125**: 1401.

Hahn, I., (2000). A contribution to consumer protection: TAD *Salmonella* vac® E – a new live vaccine for chickens against *Salmonella* Enteritidis. Lohmann Information **23**: 29-32.

Hassan, J.O. and Curtiss, R. 3rd. (1996). Effect of vaccination of hens with an avirulent strain of *Salmonella* Typhimurium on immunity of progeny challenged with wild-type *Salmonella* strains. *Infection and Immunity* **64**: 938-944.

Hinton, A.Jr., Corrier, D.E., Spates, G.E., Norman, J.O., Ziprin, R.L., Beier, R.C. and DeLoach, J.R. (1992). Biological control of *Salmonella* Typhimurium in young chickens. *Avian Diseases* **34**: 626-633.

Hinton, A.Jr., Corrier, D.E., Ziprin, R.L., Spates, G.E. and DeLoach, J.R. (1991). Comparison of the efficacy of cultures of cecal anaerobes as inocula to reduce *Salmonella* Typhimurium colonization in chicks with or without dietary lactose. *Poultry Science* **70:** 67-73.

Hodin, R. (2000). Maintaining gut homeostasis: the butyrate-NF-κB connection. *Gastroenterology* **118:** 798-801.

Holma, R., Juvonen, P., Asmawi, M.Z., Vapaatalo, H. and Korpela, R. (2002). Galacto-oligosaccharides stimulate the growth of bifidobacteria but fail to attenuate inflammation in experimental colitis in rats. *Scandinavian Journal of Gastroenterology* **37:** 1042-1047.

Hoop, R.K. and Pospischil, A. (1993). Bacteriological, serological, histological and immunohistochemical findings in laying hens with naturally acquired *Salmonella* Entertidis phage type 4 infection. Veterinary Record **133:** 391-393.

Hu, Z. and Guo, Y. (2007). Effects of dietary sodium butyrate supplementation on the intestinal morphological structure, absorptive function and gut flora in chickens. *Animal Feed Science and Technology* **132:** 240-249.

Hume, M.E., Kubena, L.F., Beier, R.C., Hinton, A.Jr., Corrier, D.E. and DeLoach, J.R. (1992). Fermentation of [14C] lactose in broiler chicks by cecal anaerobes. *Poultry Science* **71:** 1464-1470.

Humphrey, T.J. and Whitehead, A. (1993). Egg age and the growth of *Salmonella* Enteritidis PT4 in egg contents. *Epidemiology and Infection* **111:** 209-219.

Humphrey, T.J., Whitehead, A., Gawler, A.H.L., Henley, A. and Rowe, B. (1991). Numbers of *Salmonella* Entertidis in the contents of naturally contaminated hens' eggs. *Epidemiology and Infection* **106:** 489-496.

Ishihara, N., Chu, D., Akachi, S. and Jujena, L. (2000). Preventive effect of partially hydrolyzed guar gum on infection of *Salmonella* Enteritidis in young and laying hens. *Poultry Science* **79:** 689-697.

Jin, L.Z., Ho Y.W., Abdullah, N., Ali, M.A. and Jalaludin, S. (1996). Antagonistic effects of intestinal Lactobacillus isolates on pathogens of chickens. *Letters in Applied Microbiology* **23:** 67-71.

Kashket, E.R. (1987). Bioenergetics of lactic acid bacteria: cytoplasmic pH and osmotolerance. *FEMS Microbiology Reviews* **46:** 233-244.

Keller, L.H., Benson, C.E., Krotec, K. and Eckroade, R.J. (1995). *Salmonella* Enteritidis colonization of the reproductive tract and forming and freshly laid eggs of chickens. *Infection and Immunity* **63:** 2443-2449.

Kramer, J., Visscher, A.H., Wagenaar, J.A., Boonstra-Blom, A.G. and Jeurissen, S.H.M. (2001). Characterization of the innate and adaptive immunity to *Salmonella* Enteritidis PT1 infection in four broiler lines. *Veterinary Immunology and Immunopathology* **79:** 219-233.

Linde, K., Hahn, I., Vielitz; E. (1997). Development of live *Salmonella* vaccines optimally attenuated for chickens. *Lohmann Information* **20:** 23-31.

Macfarlane, G.T. and Gibson, G.R. (1996). Carbohydrate fermentation, energy transduction and gas metabolism in the human large intestine. In: (Eds. Mackie R.I., White B.R. and Isaacson R.E.) *Gastrointestinal Microbiology vol.1.* Chapman and Hall, London, U.K., pp. 269-318.

Methner, U. and Steinbach, G. (1997). Efficacy of maternal *Salmonella* antibodies against oral infection of chicks with *Salmonella* Enteritidis. *Berliner und Münchner Tierärztliche Wochenschrift* **110:** 373-377

Methner, U., Steinbach, G. and Meyer, H. (1994). Investigations on the efficacy of *Salmonella* immunization of broiler breeder birds to *Salmonella* colonization of these birds and their progeny following experimental oral infection. *Berliner und Münchner Tierärztliche Wochenschrift* **107:** 192-198

Meyer, H., Koch, H., Methner, U. and Steinbach, G. (1993). Vaccines in salmonellosis control in animals. *Zentralblatt Bakteriologie* **278:** 407-415.

Mulder, R.W.A.W., Havenaar, R. and Huis in't Veldt, J.H.J. (1997). Intervention strategies: the use of probiotics and competitive exclusion microfloras against contamination with pathogens in pigs and poultry. In: *Probiotics 2: Applications and practical aspects* (Ed. R. Fuller) Chapman & Hall, London, pp. 187-207.

Nakai, S.A. and Siebert, K.J. (2003). Validation of bacterial growth inhibition models based on molecular properties of organic acids. *International Journal of Food Microbiology* **86:** 249-255.

Nisbet, D.J., Corrier, D.E., Scanlan, C.M., Hollister, A.G., Beier, R.C. and DeLoach J.R. (1993). Effect of a defined continuous-flow-derived bacterial culture and dietary lactose on *Salmonella* Typhimurium colonization in broiler chickens. *Avian Diseases* **37:** 1017-1025.

Nisbet, D.J., Corrier, D.E., Scanlan, C.M., Hollister, A.G., Beier, R.C. and DeLoach J.R. (1994). Effect of dietary lactose and cell concentration on the ability of a continuous-flow-derived bacterial culture to control *Salmonella* cecal colonization in broiler chickens. *Poultry Science* **73:** 56-62.

Nógrady, N., Imre, A., Rychlik, I. and Barrow, P.A. (2003). Growth and colonization suppression of *Salmonella enterica* serovar Hadar in vitro and in vivo. *FEMS Microbiology Letters* **218:** 127-33.

Nurmi, E. and Rantala, M. (1973). New aspects of *Salmonella* infection in broiler production. *Nature* **241:** 210-211.

Oyofo, B.A., DeLoach, J.R., Corrier, D.E., Norman, J.O., Ziprin, R.L. and Mollenhauer, H.H. (1989). Effect of carbohydrates on *Salmonella* Typhimurium colonization in broiler chickens. *Avian Diseases* **33:** 531-534.

Pascual, M., Hugas, M., Badiola, J.I., Monfort, J.M. and Garriga, M. (1999). Lactobacillus salivarius CTC2197 prevents *Salmonella* Enteritidis colonization in chickens. *Applied and Environmental Microbiology* **65:** 4981-4986.

Rossi, M., Corradini, C., Amaretti, A., Nicolini, M., Pompei, A., Zanoni, S. and Matteuzzi, D. (2005). Fermentation of fructooligosaccharides and inulin by bifidobacteria: a comparative study of pure and fecal cultures. *Applied and Environmental Microbiology* **71:** 6150-6158.

Sadeyen, J.R., Trotereau, J., Velge, P., Marly, J., Beaumont, C., Barrow, P.A., Bumstead, N. and Lalmanach, A.C. (2004). *Salmonella* carrier state in chicken: comparison of expression of immune response genes between susceptible and resistant animals. *Microbes and Infection* **6:** 1278-1286.

Shivaprasad, H.L., Timoney, J.F., Morales, S., Lucio, B. and Baker, R.C. (1990). Pathogenesis of *Salmonella* Enteritidis infection in laying chickens. I. Studies on egg transmission,; clinical signs, fecal shedding, and serological responses. *Avian Diseases* **34:** 548-557.

Spring, P., Wenk, C., Dawson, K.A. and Newman, K.E. (2000). The effects of dietary mannanoligosaccharides on cecal parameters and the concentrations of enteric bacteria in the ceca of *Salmonella*-challenged broiler chicks. *Poultry Science* **79:** 205-211.

Springer, S., Lehmann, J., Lindner, Th., Thielebein, J., Alber, G. and Selbitz, H.-J. (2000). A new live *Salmonella* Enteritidis vaccine for chicken – experimental evidence of its safety and efficacy. *Berliner und Münchner Tierärztliche Wochenschrift* **113:** 246-252.

Thompson, J.L. and Hinton, M. (1997). Antibacterial activity of formic and propionic acids in the diet of hens on *Salmonella*s in the crop. *British Poultry Science* **38:** 59-65.

Timoney, J.F., Shivaprasad, H.L., Baker, R.C and Rowe, B. (1989). Egg transmission after infection of hens with *Salmonella* Enteritidis phage type 4. *Veterinary Record* **125:** 600-601.

Tuohy, K.M., Rouzaud, G.C., Brück, W.M. and Gibson, G.R. (2005). Modulation of the human gut microflora towards improved health using prebiotics – assessment of efficacy. *Current Pharmaceutical Design* **11:** 75-90.

Tzortzis, G., Goulas, A.K., Gee, J.M., Gibson, G.R. (2005). A novel galactooligosaccharide mixture increases the bifidobacterial population numbers in a continuous in vitro fermentation system and in the proximal colonic contents of pigs in vivo. *Journal of Nutrition* **135:** 1726-1731.

Van Coillie, E., Goris, J., Cleenwerck, I., Grijspeerdt, K., Botteldoorn, N., Van Immerseel, F., De Buck, J., Vancanneyt, M., Swings, J., Herman, L. and Heyndrickx, M. (2007). Identification of lactobacilli isolated from the cloaca of laying hens and characterization for potential use as probiotics to control *Salmonella* Enteritidis. *Journal of Applied Micriobiology* **102:** 1095-1106.

Van Immerseel, F., Boyen, F., Gantois, I., Timbermont, L., Bohez, L., Pasmans, F., Haesebrouck, F. and Ducatelle, R. (2005). Supplementation of coated butyric acid in the feed reduces colonization and shedding of *Salmonella* in poultry. *Poultry Science* **84:** 1851-1856.

Van Immerseel, F., De Buck J., Boyen F., Bohez, L., Pasmans, F., Volf, J., Sevcik, M., Rychlik, I., Haesebrouck, F. and Ducatelle, R. (2004). Medium chain fatty acids decrease colonization and invasion through hilA suppression shortly after infection of chickens with *Salmonella enterica* serovar Enteritidis. *Applied and Environmental Microbiology* **70:** 3582-3587.

Van Immerseel, F., De Buck, J., Meulemans, G., Pasmans, F., Velge, P., Bottreau, E., Haesebrouck, F. and Ducatelle, R. (2003). Invasion of *Salmonella* Enteritidis in avian intestinal epithelial cells in vitro is influenced by short-chain fatty acids. *International Journal of Food Microbiology* **85:** 237-248.

Van Immerseel, F., Methner, U., Rychlik, I., Nagy, B., Velge, P., Martin, G., Foster, N., Ducatelle, R. and Barrow, P.A. (2005). Vaccination and early protection against non host-specific *Salmonella* serotypes in poultry; exploitation of innate immunity and microbial metabolic activity. *Epidemiology and Infection* **133:** 959-978.

Waldroup, A.L., Skinner, J.T., Hierholzer, R.E. and Waldroup, P.W. (1993). An evaluation of fructooligosaccharide in diets for broiler chickens and effects on *Salmonellae* contamination of carcasses. *Poultry Science* **72:** 643-650.

Woodward, M.J., Gettinby, G., Breslin, M.F., Corkish, J.D. and Houghton, S. (2002). The efficacy of Salenvac, a *Salmonella enterica* subsp. Enterica serotype Enteritidis iron-restricted bacterin vaccine, in laying chickens. *Avian Pathology* **31:** 383-392.

The challenge of managing health in an integrated business

Dennis P. Wages
College of Veterinary Medicine, North Carolina State University, 4700 Hillsborough Street, Raleigh, North Carolina 27606, USA

1. Introduction

In the 1970's it was apparent that for poultry companies to maintain a competitive advantage in the global market they had to be able to control their own destiny. In the embryonic stages of vertical integration, companies first controlled the live production of the meat birds. Companies purchased chicks, then contracted growers or owned company farms to manage and feed out the birds. As vertical integration matured, companies took control of the parent stock breeders, hatcheries and feed mills controlling live production from breeders to processing. The last part of the integration was controlling marketing and sales of their products, which matured in the mid 1980's in the United States (US). As a result, vertical integration evolved within the poultry industry and become the rule rather than the exception. As vertical integration grew it was apparent that this type of management system could provide powerful tools not only for managing company resources but also for controlling diseases. Integrators could cut costs by buying large quantities of grain, pharmaceutical products and equipment. Likewise, with respect to diseases control, if a specific strain of pathogen produced disease in broilers, it could be prevented by maternal antibodies. The pathogen could be isolated and a killed autogenous or commercial vaccine could be produced and used in breeding flocks. This vaccine could then be used to provide maternal antibodies for the chicks and help prevent disease. In fact, this is how early cases of infectious bursal disease were controlled in the US. Likewise, as poultry integrators grew in size, pharmaceutical companies started addressing the needs of vaccinations and medications for poultry.

Within an integrated poultry company, when breeders, hatcheries, live production and processing are controlled, multifaceted disease control measures can be implemented much easier than in non-integrated poultry companies. However, integration has also led to the consolidation of poultry companies with fewer integrators producing a larger percentage of a country's processed poultry meat. The reality is that poultry is now a big business. The smaller 2-house farm is almost obsolete and now larger farms with as many as 12-24 houses are not uncommon in today's international industry. As integration and consolidation spread throughout the poultry industry, smaller poultry companies could not compete with feed and live production costs. Unless a small company has a niche market or specialty product, competing with large integrators though not impossible, is challenging.

Even with the advantages of vertical integration in the poultry industry, there are still challenges in maintaining the health status of our production flocks. Vertical integration has not been the magic bullet for disease control. It is a challenging time for the international poultry industries in terms of our disease control approaches as often we are addressing disease complexes which have to be confronted with a package of preventive tools combining measures from farm management, veterinarian and diagnostic and nutrition services. Long term strategies also have to include poultry genetics. Within the scope of this paper, it is impossible to address all of the challenges in managing health for poultry integrators but the most prudent for today's poultry integrators will be discussed. Food safety control is possible with proper integration and control over the production chain; this is another important motivator for poultry health, in an integrated business.

1.1. Food safety

Food safety, as related to human food borne illness, continues to be an international priority. In poultry, the two main bacteria of interest are *Salmonella* spp. and *Campylobacter* spp. Multiple control strategies have been utilized to decrease the incidence of bacterial contamination in fresh poultry. These strategies include: pre and post-chill processing plant strategies, biosecurity; vaccination; competitive exclusion; organic acid treatments; cleaning/disinfection; testing/depopulation; fructose and mannanoligosaccharides; antimicrobials, feed quality control; and experimental use of bacteriophages. Each of the above approaches and combinations of these strategies have had mixed results. Some poultry integrators have had very good success in eliminating *Salmonella* spp. from breeders using multiple control strategies. In spite of the poultry industry's best efforts food borne illness remains a top concern of integrators, public health officials and consumers worldwide.

Currently in the US, the Department of Agriculture holds processing plants to a *Salmonella* percentage standard. Plants are placed into categories based on percent contamination of post chill poultry carcasses: those contaminated at 20% or over are placed into category 3; 10%-20% placed into category 2; and less than 10% tolerance are placed into Category 1. If the these standards change to a 'zero tolerance' policy, as is being proposed in Europe, then prevention strategies will need to be improved. It will take an innovative, integrated and multifaceted approach to consistently reduce/eliminate *Salmonella* in fresh product. There is also concern that after *Salmonella* is addressed the same type of standards will be used to prevent *Campylobacter* spp. contamination in fresh poultry. There is already preliminary research that suggests that *Campylobacter* spp. is potentially vertically transmitted (Shanker *et al.*, 1986) which can lead to control problems both in breeders and market birds. The vertical transmission of *S. enteritidis* made control very challenging when it was first identified as a food borne pathogen.

The future of the food safety challenge in the US may be solved with irradiation, individual chilling, or combination of approaches. Control of food borne pathogens at both the breeder and grow-out level will be the rule rather than the exception. The 'farm to the table' safety concept will be important for the future of the poultry industry. This concept is where critical control points are identified, as areas where contamination may occur. Once these areas are identified, preventative measures are identified and put in place to prevent contamination from occurring. Many processing plants are already using this concept called Hazard Analysis Critical Control Point (HACCP) to minimize the potential for bacterial food contamination.

2. Managing disease without feed grade preventative antimicrobials

Europe has already experienced the challenge of raising poultry without preventative feed grade antimicrobials, with mixed results. Regardless of the reasoning behind the feed grade antimicrobial withdrawals and whether there is agreement or not for the withdrawals, the result has been a challenge for disease occurrence and approaches to prevention. This is especially true regarding bacterial enteric diseases like necrotic enteritis. In Europe as well as in the US, necrotic enteritis and enteric dysbacteriosis has been a challenge to control in antibiotic free and or organically raised chickens and turkeys. US government regulations have not, to date, removed all feed grade preventative antibiotics from poultry production. However, consumer pressure is strongly encouraging poultry companies to remove disease prevention feed grade antimicrobials. In reality, major grocery food chains, fast food restaurants and other food retail companies play more of an important role in production decisions, than the live production managers themselves. Therefore, those that know the least about poultry production are actually the driving force on how poultry integrators make day to day decisions on live production and disease prevention guidelines.

Today in the US, virginiamycin, and lincomycin; two common feed grade necrotic enteritis control drugs are in jeopardy of being pulled from the market. Virginiamycin is a streptogramin and is under pressure for withdrawal because of new streptogramin products being cleared for use in humans to treat gram positive nosocomial infections. In the case of lincomycin, its close relation to the human drug clindamycin has resulted in public health outcries for its removal. Every year in the US Congress, a bill is introduced into the government legislature to ban all feed grade antimicrobials that are either used in humans or have a product related that is used treat human infections. However, as an industry we either have failed at educating our customers of the benefit and safety of our antimicrobials that we use in food animals, or we are just simply outmanned, out financed and do not have the resources to carry the message to our customers.

There is reason for optimism, however, product approval and research on novel non-antibiotic agents are increasing exponentially. Mannanoligosaccharides, essential oils, plant extracts and other natural products are providing exciting opportunities for disease control, especially in antibiotic free and organic raised poultry. In fact over the last 3-5 years there has been an impressive increase in their use, with successful results. Research has shown that these products not only have antimicrobial and antiprotozoal effects but also prevent the attachment of certain pathogenic enteric bacteria such as *Salmonella* spp. and *E. coli*. Mannanoligosaccharides have been shown to; increase jejunum length, increase key intestinal enzyme activities (Lji *et al.*, 2001) and compliment coccidial vaccination (Nollet *et al.*, 2007). They have also been found to increase the concentration of beneficial bacteria such as lactobacilli and bifidobacteria (Baurhoo *et al.*, 2007), as well as enhancing production parameters and immunity (Shashidhara and Devegowda, 2003).

Vaccines for common bacterial diseases are also being actively researched. *Escherichia coli* vaccines, both live and killed, targeting O and K antigens have been produced with varying results (Barnes *et al.* 2003). Likewise, a *Clostridium perfringens* type A toxoid used in broiler breeders has also been used successfully, to prevent both clinical and sub-clinical necrotic enteritis. Organic and antibiotic free chicken production commonly uses this toxoid. Vaccines for fowl cholera both live and killed are also used with success.

Another important way to address the challenge of managing disease without preventative medication is the use of diagnostics. Most of the preventative medications utilized in the US are to prevent bacterial enteric disease. However, many of the bacterial enteric diseases such as *Clostridia* and *E. coli* infections are usually secondary to a viral and or protozoal infection. A diagnostic work up for enteric disease is essential to implement an effective preventative enteric disease program.

2.1. Case study

The following discussion relates to troubleshooting and managing diagnostics in an enteric disease of young turkeys: the most common challenge in the US turkey industry today. This occurs in the young turkey poult between the ages of 7 days and 35 days of age but most commonly observed in the first 10-24 days of age. This condition is commonly called a poult enteritis complex. The condition is called like this because multiple aetiologies are usually involved in this complex disease. This condition is more often referred to as poult enteritis mortality syndrome (PEMS) referring to the amount of mortality observe in flocks. Multiple etiologies involved in poult enteritis complex are: viruses (such as coronavirus, rotavirus, astrovirus, and enterovirus); protozoa (i.e. coccidia species, *Cochlosoma*, *Spironucleus*, *Trichomonas*, and *Cryptosporidium*) and bacteria (*E. coli*, *Clostridia* spp.) Most commonly viruses set up the initial infection and compromise the intestine or change intestinal pH to allow other opportunistic organisms to complicate the viral insult. The clinical signs are so similar for the different aetiologies involved with poult enteritis complex that a complete diagnostic work up is essential. To accurately diagnose an enteric disease in turkey poults the first place to start is performing a thorough necropsy on 5-10 typically affected birds to identify any gross lesions. If the wall of the upper small intestine is thin and the intestinal content is watery, a viral aetiology should be suspected. If the content has focal areas of necrosis or increased fibrin, a clostridial infection (clinical or sub clinical necrotic enteritis) should be suspected. However, one must remember that often multiple aetiologies are involved so the diagnostic work up must not stop at gross lesions. After gross inspection, intestinal smears are taken from the duodenum, mid jejunum, mid ileum and cecum. They are evaluated under a relatively low 25X power microscope to determine the amount of bacteria observed or if any protozoa are identified. A composite faecal samples is submitted for electron microscopy for enteric viruses and is stained with Auromine-O, to determine if *Cryptosporidium* is present. Sections of the duodenum loop, jejunum and the ileo-cecal junction, are fixed in formalin to see if the microscopic lesions correlate with the gross lesions, intestinal smears, faecal examination and virus identification.

2.2. Diagnostics, treatment and prevention

Often the viral tests by electron microscopy are negative because the virus has already been eliminated with intestinal epithelia cell sloughing. Microscopic evaluation of the intestines is a very important component for the complete diagnostic work up. There are some enteric viruses that can be identified by fluorescent antibody testing or immunofluorescent in fresh intestines. For enteric viruses like coronavirus, serology can also be performed.

The diagnostic work-up for enteric disease is essential because different aetiologies may be approached quite differently for treatment or control. Coccidia infections would result in the investigation of the coccidiostat used or if coccidia vaccination was properly performed. When protozoa other than coccidia are involved, control may be more problematic. In the US, organic arsenicals in the feed are still approved for use for prevention of protozoa infections and have some efficacy. When these become unavailable, new, novel, anti-protozoa feed grade medication will be essential in both broilers (Hume *et al.*, 2006) and turkeys.

Likewise, a bacterial infection could be prevented with the addition of a mannan oligosaccharide, a probiotic, or antimicrobials if effective and available (Fairchild *et al.*, 2001). When viruses are suspected, usually cleaning and disinfection is warranted as well as strict farm biosecurity. As with most enteric viruses in poultry, older birds that have recovered remain infected carriers and serve as a source of infection to the young birds, especially on a two-stage or multi-age farm. Insuring that the virus is not tracked back into the younger birds is an important control tool.

3. Managing disease with a limited number of therapeutic antimicrobials

Therapeutic antimicrobials have diminished to the point where treating flocks is no longer a favourable option for disease treatment, for many bacterial and protozoal infections. Not only does this affect live production parameters and costs, it creates an animal welfare issue if there is not an efficacious antimicrobial to treat sick flocks, in order to minimize suffering, pain and distress. For example, in the 1970's histomoniasis (black head) was controlled and treated with dimetridazole and ipronidazole. Since their removal, we have no therapeutic options for black head outbreaks. Instead, we have to rely on organic arsenicals in the feed and water to minimize mortality. The organic arsenicals are currently under scrutiny in the US because of residues and their environmental impact, so that their subsequent future availability is questionable. Also, the performance of the arsenicals in an acute outbreak of black head is fair at best. In the 1980's, nitrofurazone (NFZ) was an excellent treatment for *E. coli* airsacculitis, after the removal of NFZ, there was no effective antimicrobial for treatment of airsacculitis in broilers. Poultry integrators waited approximately 6 years for the approval of enrofloxacin to treat fowl cholera in turkeys, as well as colibacillosis in both turkeys and chickens. In 2004, this product was also removed from the market for use in poultry which left the poultry industry severely impaired in treating colibacillosis. The tetracycline class of antimicrobials is currently the only approved products for *E. coli* airsacculitis and most *E. coli* isolates demonstrate greater than 90% resistance. The sulfonamide drug class is used in turkeys with success for fowl cholera and *E. coli* control however, because of the short production cycle in broilers and the potential for residues, sulfonamides are not used in broilers. On 1st October 2008, a cephalosporin product ceftiofur will no longer be able to be used in an extra label manner for food animals, in the US. Its current use in poultry as an in-ovo injection combined with Marek's disease will no longer be allowed.

Veterinarians in US integrated poultry companies can prescribe antimicrobials for therapeutic use in water or by injection only under the Animal Medicinal Drug Use Clarification Act of 1996 (Food and Drug Administration, 1998), also referred to as the Extra Label Drug Use Act. However drugs available in water soluble formulation for use in poultry flocks are either not manufactured anymore, limited in number, ineffective or the products are cost prohibitive. Currently in the US there are only 10 antimicrobials that can be used in water for treating poultry and all were developed decades ago. Companion animal antimicrobials have limited or no use in poultry because of their formulation and many of them are illegal to use in food animals. From a feed additive standpoint, US veterinarians cannot prescribe off label use of antimicrobials in feed if they are not approved and listed in the Feed Additive Compendium (Animal Health Institute, 2007). That is, if a veterinarian would like to use Antimicrobial A at 600 grams per ton and the approved, listed levels in the Feed Additive Compendium are in the range of 200-400 grams per ton of feed, the veterinarian is restricted to what is listed in the Feed Additive Compendium. Likewise, if a veterinarian would like to prevent the disease at 100 grams per ton they would be prohibited to do so.

As discussed earlier with preventative medication, diagnostics play an important role in preventing disease where a therapeutic agent is warranted. It is important that the approach is well designed and leaves minimal room for an infection to go unidentified. In broiler chickens, colibacillosis serves a good example on how to address the issue. The most common condition that predisposes the broiler to *E. coli* infection is respiratory disease. This condition is mostly seen as a late breaking (>5 weeks of age) airsacculitis.

3.1. Case study

Presented below is an example of how a diagnostic approach can be used, in this case for broilers with respiratory colibacillosis. After confirming the colibacillosis and identifying the farm involved, specific pathogen free sentinel (SPAFAS) chickens are placed in the house when the broilers are 4-5 weeks of age. The timing of sentinel placement is at least two weeks after any field boost of Newcastle (NDV) or infectious bronchitis (IBV) vaccine has been performed. Seven days after placement, two tracheal swabs are taken per sentinel and this is repeated a further seven days later. One of the tracheal swabs is cultured on both blood and MacConkey agar for bacterial isolation such as *Ornithobacterium*. Then that swab is placed in Frey's media for transport to the laboratory and mycoplasma isolation. The second swab is placed in a virus support broth solution with antibiotics added for virus isolation. The virus isolation samples should always be ice packed or refrigerated until shipping. At 14 days post sentinel placement or at the time of slaughter, whichever is longer, blood is collected from each sentinel, the serum separated and refrigerated for later analysis. At the same time, the sentinels are sacrificed and a complete necropsy performed. The lower two thirds of the trachea is collected and submitted to a diagnostic laboratory for virus and mycoplasma culture. The upper portion of the trachea is placed in formalin for microscopic evaluation. Next the caudal thoracic air sacs are collected, cultured for mycoplasma and placed on ice for virus isolation. If the broiler flock has not gone for processing yet, 20 blood samples are collected, the serum separated and refrigerate for analysis.

3.1.1 Diagnostics, treatment and prevention

An enzyme linked immunosorbent assay (ELISA) is carried out on the serum, for NDV, IBV and mycoplasma. Hemagglutination inhibition (HI) tests can also be carried out for NDV and mycoplasma. If IBV is isolated, all isolates are classified by strain to determine if an antigenic shift has occurred but this is much less common with NDV. Based on isolations, appropriate preventative measures can be implemented. One of the most difficult things to control is when a variant or antigenic shift has occurred with IBV and no cross protection with current commercial vaccines is provided. Usually clean out, disinfection and down time is the only effective preventative alternative. When NDV is identified as the problem, usually changing the dose of NDV spray at the hatchery or reviewing the NDV booster type and route, usually helps to control NDV. Since most mycoplasma infections in broilers are controlled by eradication in the US, control measures in broilers are not common. Whenever respiratory colibacillosis is confirmed in a flock, immunosuppressive agents such as infectious bursal disease, chicken anaemia virus and mycotoxins, should be ruled out.

4. Biosecurity challenges

Biosecurity is a word that is used quite extensively when poultry disease outbreaks occur. Particularly if those outbreaks are diseases such as avian influenza (AI) or velogenic viscerotropic Newcastle disease (VVND), due to their devastating effect on the poultry industry. With respect to H5N1 AI, if the potential human disease risk is also considered then this heightens awareness and brings public health officials into the discussions. However, diseases such as laryngotracheitis, mycoplasmosis, and pullorum bring out similar discussions regarding biosecurity and are much more common. There are numerous definitions of biosecurity. Wikipedia Encyclopedia defines biosecurity as 'the policies and measures taken to protect from biological harm. It encompasses the prevention and mitigation from diseases, pests, and bioterrorism, of the following areas: economy, environment, and public health. This includes food and water supply, agricultural resources and production, pollution management, blood and blood product supplies'. In simpler terms biosecurity for poultry production is 'the management measures taken to prevent the introduction of disease onto a poultry premise, if disease is diagnosed to keep it there and let it die there'. There is little that virtually all poultry integrators have some type of biosecurity policy or guidelines in place, which are usually comprehensive. Factors addressed in guidelines include: traffic control; visitor restriction; depopulation; education on disease spread, pest/insect/rodent/animal control; avoiding contact with non-commercial poultry and wild birds; premise cleanliness; boots, hair net and coverall use; live bird markets; dead bird disposal; do not trade or exchange equipment with other growers; cleaning and disinfection; feed truck options; live haul restrictions; litter disposal; vaccination; communications; bird movement; and public utilities access (LP gas, propane, etc). However, when discussing biosecurity in parts of the world, where cock fighting, open city markets with poultry and poultry live markets are common, biosecurity is a challenge. When farms that keep 25-100 non-commercial poultry, take those animals to markets in a wide geographical area, it is very difficult to prevent disease spread. A critical part of the small, non-commercial poultry farms is that some of the people involved may actually work for the poultry integrator, with out them being aware of it. For the poultry integrator, multi-age farms with 15-25 houses are a biosecurity nightmare. It is next to impossible to place all of the birds at the same time or even within the same week on multi-age large farms. As such, traffic control within those farms is extremely difficult.

The challenge for the poultry integrator is 'biosecurity compliance'. How can they be sure that the live haulers, growers, plant employees and other company employees are doing what they need to do to prevent the introduction and spread of disease? It is one belief that poultry integrators should have biosecurity auditors, whose sole responsibility is insuring that the biosecurity plans are being adhered to. These individuals would become most valuable when an exotic or reportable disease is introduced into an area.

However, biosecurity's biggest hurdle may be convincing upper management in poultry integrator companies that biosecurity guidelines are not only essential but also cost effective. It is very difficult to prove the cost effectiveness of an event that may never happen. Spending millions of dollars to prevent the introduction of a catastrophic disease is a hard sell to upper management. However, when biosecurity is discussed with individuals or companies that have been involved with the current H5N1 AI outbreak, cost effectiveness and compliance will be supported. The 1983-84 the H5N2 AI outbreak in Pennsylvania cost the industry 65 million US dollars to control and eradicate, directly impacting 6 companies. The cost of that AI eradication would have paid for 20.3 million pairs of coveralls or 135.4 million pairs of disposable boots or 940,000 gallons of disinfectant. How many of those 6 companies would have approved over 10 million dollars each to establish and implement their biosecurity programs?

A major component of biosecurity is that of depopulation of flocks. To date, the world has been lucky that a catastrophic disease has not escaped and affected large poultry integrators in several countries, which would impact on hundreds of millions of birds. In many countries the procedures for such an event are not properly planned and therefore its implementation would be very difficult and pose additional risks.

Biosecurity cannot be discussed without considering the lack of continuity within the poultry industry, which is no longer cohesive or unified with respect to disease control. When discussing major biosecurity issues involving a large geographical area, representatives from the following industries must be at the table: broilers, primary broiler breeders, turkeys, primary turkey breeders, table egg layers, live poultry markets, organic, antibiotic free and other niche poultry markets. The different parts of the poultry industry are diverse in their needs and thinking, with respect to biosecurity and disease control. What is acceptable to a large broiler or turkey integrator may not be acceptable to a small organic producer supplying a niche market. In the US it may take government intervention with mandatory compliance to ensure a poultry industry consistent in biosecurity and disease control programs. However, excessive government involvement is not always welcomed and to date it has not been very successful in addressing all facets of the poultry industry in the US or worldwide.

The poultry industry is at a turning point, where every poultry integrator or small poultry company and anyone that manages commercial and non-commercial poultry, must assume avian influenza or some other catastrophic disease is either on their farm or only 50 meters away. Everything possible must be done to keep it off farms or if already on a farm, to ensure that it never leaves it.

5. Vaccination / immunization

Vaccination has been one of the areas where poultry integrators have made great strides. Vaccination of breeders to protect health, egg production and the progeny through maternal antibodies transfer is unparalleled in the other food animal industries. The use of in-ovo vaccination has proven to be a tremendous benefit to poultry integrators. Originally used for Marek's Disease control, this technology is now being used to vaccinate for laryngotracheitis, fowl pox and infectious bursal disease (IBDV). This novel delivery system is also an essential component in the use of vectored vaccines. Future multivalent and vectored in-ovo vaccination may well be the rule rather than the exception for disease control in the next decade (Sharma *et al.* 2002). The poultry industry's innovation of hyper-immunizing breeder hens to control diseases such as IBDV, is now one of the most common means to control IBDV in broiler chickens. Research on and the use of recombinant and vectored vaccines continues to show promising results and will most likely be the future for vaccination of poultry.

However, there are some challenges in the area of vaccination that I believe warrants further discussion. Antigenic drift and the development of variant virus strains, in diseases that we deal with on a daily basis such as infectious bronchitis and infectious bursal disease, are a continuous challenge. For example, infectious bronchitis (IBV) has historically been a challenge to control through vaccinations. Some of the original IBV strains used in the US, such as the Massachusetts and Connecticut strains used for years to control this disease, are still in use and are efficacious. However, Delaware O72 and Arkansas strains of IBV continue to produce antigenic drifts or variants, so that current vaccination control is not as effective as in the past. International IBV variants are named based the number of passages on cell culture or embryo. Pharmaceutical companies can no longer afford to produce live IBV vaccines for variant IBV viruses. By the time research, development, and clinical trials are finished the vaccine will no longer be effective or its efficacy diminished. The time has come to re-think the extensive use of live vaccines in poultry or at least re-think the route of delivery for some of the live vaccines such as IBV. In the United Kingdom, research is being carried out to vaccinate IBV in-ovo through the use of an IBV hybrid virus.

The poultry industry's approach to disease control will most likely involve molecular approaches. The discovery of cytosine phosphoric acid and guanidine (CpG) and their use to prevent *E. coli* cellulitis infection experimentally through immunomodulation is the innovative approach to disease control (Babiuk *et al.*, 2003). Immune complex vaccine use will continue to increase and play a vital role in future disease prevention in both chickens and turkeys. The continued development of novel, innovative, safe, cost effective disease control programs should continue to be the focus of research and development in the next few years. Once the new technology presents novel vaccines we should concentrate our research on the traceability of vaccine strains and being able to differentiate vaccine strains from field challenge strains. Diseases that historically have been controlled by eradication and depopulation may, in fact be controlled by vaccination through novel vaccines. Future vaccine technology may have the possibility to control AI or mycoplasma in broiler and turkey breeding flocks. The future potential in vaccine and vaccination technology is tremendous and will continue to play a pivotal role in disease prevention.

6. Genetics

For years, production parameters and more specifically live weight were the focus of breeding programmes, especially in chickens. Often industry customers have driven this genetic focus. As genomics has matured over the past 10-20 years, egg production, breast yield and feed conversion, have been a major focus for broiler and layer breeding. The end result at the processing plant is driving much of the genetic research and it may be at the expense of the breeder. Genetic research for meat quality, yield, cardiovascular improvement and musculoskeletal traits in broilers are a high priority for breeders because it is a high priority in the finished product. The mortality from ascites and sudden death or 'flip over' associated with fast rates of growth is still unacceptable. However, the challenge for breeders is to provide a bird which performs through the breeding and growing phases and at the processing plant, including disease resistance to respiratory and enteric disease. Research to recognize the genetic control of disease will be paramount in future disease control for integrators. The identification of the major histocompatibility complex (MHC) and its role in immunity to Marek's disease (Lamont, 1998) is just one example where genetics has played a vital role in disease control.

7. Feed quality

Feed has to provide all nutrients to assure that the requirements of the birds are covered, it must ensure optimal gut development and function and should contain minimal quantities of compounds which negatively affect the bird health and performance. The most important compounds which can negatively impact the bird are mycotoxins. Mycotoxins are thought and many times documented to be involved in the development of: tibial dyschondroplasia, cage layer fatigue, rickets, malabsorption, capillary bleeding, immunosuppression, femoral head necrosis, articular gout, and foot pad dermatitis. The *Fusarium* class of mycotoxins is most commonly incriminated in the above clinical syndromes. However, aflatoxins and other classes of mycotoxins are interrelated within these conditions. Musculoskeletal problems resulting in culling, decrease weight gain, poor feed conversion, and carcass condemnations in both chickens and turkeys has been estimated to be over 2.5 billion US dollars during the 1990's. A large portion of these problems have been associated with mycotoxin contamination of feed. In the US, floods, droughts and the diversion of feed grain corn to ethanol production has affected 30% of the corn production in the Midwest. This has driven the cost of feed to unprecedented levels and has affected the quality of grain that is available for livestock feed. Mycotoxin prevention continues to be a challenge and a priority for poultry integrators.

8. Welfare impact on health

Unfortunately, there are more questions than answers on this subject because it is difficult to predict the impact animal welfare regulations will have on poultry health. Most people not directly involved in poultry production believe in the importance of animal welfare for poultry. With available space and outdoor access being key issues, the public often defines welfare differently from industry experts. Poultry integrators and veterinarians understand that unless birds are well fed, well managed and disease free (raised under best management practices), they will not perform to optimal standards. The question is, how much welfare is enough and what shall be the key parameters? Looking into the future reflects unpredictability. It is apparent that important customers (not consumers) will alter our management of birds. For example, in the US a major fast food provider halted the moulting of table egg layers in most layer companies because of outside organization (animal rights) pressure. In Europe the historic table egg layer caged system will change because of outside pressure. This is not only occurring in poultry but also in the swine and cattle. Animal welfare policy may shape the future of our industry more than any other guideline, policy or rule.

Everyone agrees that best management practices and animal welfare guidelines in poultry result in optimal productivity and health. However many additional issues are today and will in the future be discussed much more controversially:
- Will we be able to raise pullets under restricted lighting?
- Will all animals need access to the outside?
- Will there be no caged layers as we know it today?
- Will we have to have a minimum amount of light to raise meat birds?
- Will a certain incidence of diseases in a company shut it down?
- Will we have to clean out litter after every flock in the US?
- Will lameness or gait scoring be mandated?
- Will welfare regulations tell us when to treat?
- How about the diseases where we have no treatment approved in the US; will we be forced to depopulate that farm and who will pay for it?

Welfare regulations and auditing are evolving and many would agree most of the regulations are appropriate. However, what will be the consequences for failure and who will set those consequences?

9. Conclusion

Overall, poultry integrators are doing a very good job in providing a consistently high quality, safe protein product at an affordable price and should be commended. Poultry companies continue to improve productivity while lowering disease challenge and reducing condemnations in the plant. The similarities and differences within the poultry industry must be understood if an integrated approach to disease control it to be workable. We must recognize that the cost and management involved in raising poultry in the Ukraine is not the same as in France. Likewise, the cost of production and management needed in raising poultry in South Texas cannot be compared to Maryland. The poultry industry, even though similar in many ways, has a wide range of diversified practices within it. Every poultry company collects reams of information on their flocks from the breeding through the feeding and processing. The enormous amount of data collected by companies and then examined by individuals, universities and agricultural statistics companies must be better used, in order for integrators to make informed decisions and not used for punitive company actions.

The poultry industry is facing a paradigm shift in its thinking and approach to disease control. Innovative and novel approaches to disease prevention, treatment and control are more important for the future than ever before. The idea of least cost production may need to be redefined as 'effective least cost production'. Some cost cutting in fact cost the industry more, a hard look at the way vaccine dosages are cut and biosecurity guidelines are short circuited. There is also a tendency to blame problems which are not understand on the breed, feed or management.

Fuel and grain prices worldwide will continue to drive our cost of poultry production higher. The antimicrobial arsenal for prevention and control of bacterial and protozoal diseases will continue to be reduced. Novel, natural and innovative products must take their place. Diagnostics should be more widely used to help us prevent diseases that infect or predispose poultry to common and catastrophic infections. Animal welfare of poultry may suffer because of a lack of therapeutic antimicrobials. After diagnosis, new and novel vaccination must be ready for use. Biosecurity is not just a word to satisfy upper management and government officials, but is hard work and those involved in those policies must be held accountable. The poultry integrator and upper management must understand, with respect to biosecurity, that it is difficult to put a cost on a program that keeps a catastrophic disease out of poultry integrations.

Genetics companies need to continue to focus on providing the complete package including disease resistance for poultry integrators. It is likely that in the future animal welfare aspects will play a more important role in animal genetics. Many male broilers have abnormal gaits due to the size of the breast muscle and this issue is also seen in male turkeys. Rapid growth rate is not an advantage if the result is late grow-out mortality. Some viral challenges are not best defended by live vaccines, vaccine and immune technology must continue to provide innovative and novel products that are effective at a reasonable cost. The complete package will continue to be priority. 'We can no longer continue do what we do just because we have always done it that way'.

References

Animal Health Institute (2007). *Feed Additive Compendium*. Miller Publishing Company, Minnetonka, Minnesota, USA.

Babiuk, L.A., Gomis, S. and Heckert, R. (2003). Molecular approaches to disease control. *Poultry Science* **82**: 870-875.

Barnes, H.J., Vaillancourt, J.P. and Gross, W.B. (2003). Colibacillosis. In: *Diseases of Poultry 11*[th] *Edition*. (Eds. Saif, Y.M., Barnes, H.J., Glisson, J.R., Fadley, A.M., McDougald, L.R. and Swayne, D.E.) Iowa State University Press, Ames, Iowa, USA, pp. 631-656.

Baurhoo, B., Phillip, L. and Ruiz-Feria, C.A. (2007). Effects of purified lignin and mannan oligosaccharides on intestinal integrity and microbial populations in the ceca and litter of broiler chicks. *Poultry Science* **86**:1070-1078.

Fairchild, A.S., Grimes J.L., Jones, F.T., Wineland, M.J., Edens, F.W. and Sefton, A.E. (2001). Effects of hen age, Bio-Mos® and Flavomycin® on poult susceptibility to oral *Escherichia coli* challenge. *Poultry Science* **80**:562-571.

Food and Drug Administration (1998). *Animal medicinal drug use clarification act*. Federal Register, USA, pp. 57731-57746.

Hume, M.E., Clemente-Hernandez, S. and Oviedo-Rondon, E.O. (2006). Effects of feed additives and mixed *Eimeria* species infection on microbial ecology in broilers. *Journal of Poultry Science* **85**: 2106-2111.

Lamont, S.J. (1998). Impacts of genetics on disease resistance. *Poultry Science* **77**: 1111-1118.

Lji, P.A., Saki, A.A. and Tivey, D.R. (2001). Intestinal structure and function of broiler chickens on diets supplemented with a mannan oligosaccharide. *Journal of the Science of Food and Agriculture* **81**: 1186-1192.

Nollet, L., Huyghebaert, G. and Spring, P. (2007). Effect of dietary mannan oligosaccharide on live performance of broiler chickens given an anticoccidial vaccine followed by a mild coccidial challenge. *Journal of Applied Poultry Research* **16:** 397-403.

Shanker, S., Lee, A. and Sorrell, T.C. (1986). *Campylobacter jejuni* in broilers: the role of vertical transmission. *Journal of Hygiene* **96:** 153-159.

Sharma, J.M., Zhang, Y., Jensen, D., Rautenschlein, S. and Yeh, H.Y. (2002). Field trial in commercial broilers with a multivalent in ovo vaccine comprising a mixture of live viral vaccines against Marek's Disease, infectious bursal disease, Newcastle Disease, and fowl pox. *Avian Diseases* **46:** 613-622.

Shashidhara, R.G. and Devegowda, G. (2003). Effect of dietary mannan oligosaccharide on broiler breeder production traits and immunity. *Poultry Science* **82:** 1319-1325.

Mycotoxins in poultry production: impact on animal performance and immunity

Jean-Denis Bailly
Mycotoxicology, National Veterinary school, 23 Chemin des caplles, BP 87614, 31076 Toulouse cedex, France

1. Introduction

Mycotoxins are a heterogeneous group of secondary metabolites elaborated by fungi. More than 300 secondary metabolites have been identified; around 30 are of real concern for human and animal health (Bennett and Klich, 2003). Mycotoxins are produced during mould development. They can be found as natural contaminants of many vegetal foods or feeds; mainly cereals but also fruits, nuts, grains, forage as well as compound foods intended for human or animal consumption. The most important mycotoxins are produced by moulds belonging to *Aspergillus*, *Penicillium* and *Fusarium* genus (Table 1) (Bhatnagar *et al.*, 2002; Pitt, 2002; Conkova *et al.*, 2003). These molecules are usually classified depending on the fungal species that produce them.

Mycotoxin toxicity is variable. Some have a hepatotoxicity (aflatoxins), others have an estrogenic potential (zearalenone) and some are immunotoxic (trichothecenes, fumonisins) (Bennett and Klich, 2003). Some mycotoxins are considered as carcinogenic or suspected to have carcinogenic properties (IARC, 1993). However, a few can have positive effects such as those which have been used to produce the antibiotic penicillin.

Due to the important structural diversity of mycotoxins, the variations in their metabolism and subsequent toxicity, it is impossible to produce a general set of rules; each toxin has to be investigated individually. Therefore, the presentation of data available on the most important toxins will be made toxin by toxin, presenting successively the origin; the impact of each molecule on animal performance and immunity. Taking into account their breeding and feeding conditions, poultry species can be exposed to many different mycotoxins. Many studies have consequently aimed to characterize the potential toxicity of these fungal metabolites in different avian species.

Table 1. Mycotoxins and producing fungal species associated with human and animal nutrition.

Mycotoxins	Main producing fungal species
Aflatoxins B$_1$, B$_2$, G$_1$, G$_2$	*Aspergillus flavus, A. parasiticus, A. nomius*
Ochratoxin A	*Penicillium verrucosum, Aspergillus ochraceus, A. carbonarius*
Fumonisins B$_1$, B$_2$, B$_3$	*Fusarium verticillioides, F. proliferatum*
Trichothecenes	*Fusarium graminearum, F. culmorum, F. sporotrichioides, F. poae, F. tricinctum, F. acuminatum*
Zearalenone	*Fusarium graminearum, F. culmorum, F. crookwellense*
Patulin	*Penicillium expansum, Aspergillus clavatus, Byssochlamys nivea*
Ergot alcaloïds	*Claviceps purpurea, C. paspali, C. africana*
Citrinin	*Aspergillus terreus, A. carneus, A. niveus, Penicillium verrucosum, P. citrinum, P. expansum*
Cyclopiazonic acid	*Aspergillus flavus, A. versicolor, A. tamarii, Penicillium camemberti*
Sterigmatocystin	*Aspergillus flavus, A. versicolor, A. nidulans*
Sporidesmins	*Pythomyces chartarum*
Stachybotryotoxins	*Stachybotrys chartarum, S. atra*
Endophyte toxins	*Neotyphodium coenophialum, N. nolii*
Tremorgenic toxins	*Penicillium roqueforti, P. crustosum, P. puberrelum, Aspergillus clavatus, A. fumigatus*

2. Impact of mycotoxins on poultry

2.1. Aflatoxins

Aflatoxins are probably the most studied and well documented mycotoxins. They were discovered following a toxic accident in turkeys fed a groundnut oilcake supplemented diet (Turkey X disease) (Nesbitt *et al.*, 1962).

2.1.1 Origin

The four natural aflatoxins (B$_1$, B$_2$, G$_1$ and G$_2$) can be produced by strains of fungal species belonging to *Aspergillus* genus; mainly *Aspergillus flavus* and *Aspergillus parasiticus* (Rapper and Fennel, 1963; Klich and Pitt, 1968). Worldwide these are common contaminants of wide variety of feed ingredients. Aflatoxins can be found in many vegetal products, from cereals to groundnuts, cotton seeds, dry fruits and spices. These fungal species can grow and produce toxins in the field or during storage and climatic conditions required for their development are often associated with tropical areas (high humidity of the air, temperature ranging from 25 to 40 °C) (Thompson and Henke, 2000; Kaaya *et al.*, 2006). When present in a vegetal matrix, aflatoxins are very stable and weakly sensitive to thermal treatments (sterilisation or freezing) or drying (Park, 2002).

2.1.2 Impact on poultry performance

As is true for most animal species, the main target of aflatoxins in poultry is the liver. Clinical signs differ depending on the dose and the time of exposure. Aflatoxin toxicity also varies as a function of the species and the age of the animals (Klein *et al.*, 2002). One-day-old ducklings being the most sensitive and adult quails the most resistant. Differences in hepatic biotransformation of the toxin could be responsible for these differences among avian species (Lozano *et al.*, 2006).

Avian species are quite resistant to aflatoxins and acute toxicity is not usually observed under breeding conditions. Chronic toxicity of aflatoxin, however, is more common in poultry. It normally follows ingestion of contaminated feeds over a period of several weeks (minimum 1 week). Clinical signs are mainly a decrease in the breeding performance of the birds as well as a decrease in the body weight gain and egg production. This is associated with haemorrhage and pigmentation of the carcass (Dalvi, 1986) and hepatic lesions are the most characteristic sign. Nodular hyperplasia with fibrosis and proliferation of the biliary ducts has been observed in duck, turkey and chicken (Newberne and Butler, 1969). These lesions were reported for doses of about 0.1 mg/kg of feed for ducks, 0.3 to 0.5 mg/kg of feed for turkeys and 0.5 to 2 mg/kg of feed for chickens. Biochemical and haematological alterations often occur during exposure to aflatoxin. A decrease in the serum concentration of proteins, cholesterol and triglycerides has been observed; as well as an increase in γglutamyl transferase, alkaline phosphatases, sorbitol-dehydrogenase and transaminases (Smith *et al.*, 1992).

Hamilton demonstrated that these toxins may also alter animal performance after exposure to weaker doses than those reported before. Indeed, he was the first to demonstrate an alteration in breeding performance during exposure to low concentrations of toxins (Hamilton, 1971, 1975). A dose of 1250 μg aflatoxin/kg of feed was required to cause a reduction in the growth rate. However, haematological and biochemical modifications were observed at levels from 625 µg/kg of feed. A more detailed analysis of low contamination levels on feed conversion ratio (significant effect for 1% variation) and on growth (significant effect for 2% variation) revealed that feed contaminated with 30 µg/kg resulted in financial losses, in 30% of the cases. Therefore, Hamilton recommended that poultry feed should not exceed 10 µg aflatoxin/kg of feed (Hamilton, 1975).

2.1.3 Impact on immunity

Aflatoxin B_1 (AFB$_1$) induces immunosuppression in animals. However, at least in poultry, the effects that have been observed are contradictory and difficult to interpret (Qureshi *et al.*, 1998). They may vary with the dose (immunosuppression usually accompanies exposure to high concentrations of toxins), the period of exposure to the infectious agent (during or after exposure to aflatoxins) and the type of immune response (humoral or cellular) (Meissonnier *et al.*, 2006).

Several studies aimed to characterize the effects of aflatoxin B_1 on lymphoid organs. Exposition of broiler chickens to high doses of AFB$_1$ (1 to 2.5 mg/kg feed) for 2 to 6 weeks leads to gross and histological lesions with lymphoid depletion in spleen, thymus and bursa of Fabricius follicles (Ortatatli *et al.*, 2001; Shivachandra *et al.*, 2003; Karaman *et al.*, 2005). High concentrations of aflatoxins also decreased the production of antibodies during immunization with sheep red blood cells (SRBC), this was more pronounced for IgG than for IgM (Virdi *et al.*, 1989). Practical consequences of AFB1 effects on immune function could be an increased susceptibility of exposed birds to infections. Indeed, experimental exposure to 5 mg AFB1/kg of feed along with *Salmonella gallinarum* infection, had an additive effect on body weight gain in chickens (Smith *et al.*, 1986). Aflatoxin B_1 exposure following natural infection with *Salmonella worhtington* also reduced body weight gain (Wyatt *et al.*, 1975). In this later experiment a synergistic effect was observed, suggesting that such effect may be seen under field conditions.

However, exposure to low concentrations of aflatoxin B_1 did not have a marked effect on immunity. For instance, in chickens, dietary exposure to 0.1 to 0.8 mg AFB$_1$/kg feed did not change the level of specific antibody, during vaccination against Newcastle disease and *Pasteurella multocida* (Giambrone *et al.*, 1985). Similarly, no modification of antibody production was observed after cholera vaccination in ducks exposed to 0.2 mg AFB$_1$/kg of feed for 3 weeks (Cheng *et al.*, 2001). By contrast, a marked effect on cell-mediated immune response could be observed after exposure to comparable doses of AFB_1. For example, a significant decrease in total lymphocytes and T cells counts was observed in broiler chickens fed for 6 weeks with 0.1 to 1 mg AFB_1/kg of feed (Ghosh *et al.*, 1991).

2.2. Ochratoxin A

In poultry, intoxication with ochratoxin A (OTA) can lead to acute or chronic toxicity. These two clinical forms are usually observed after ingestion of relatively high doses of OTA, several mg/kg of feed. Few data are available concerning the effects following exposure to low doses, only µg/kg of feed.

2.2.1 Origin

Ochratoxin A was isolated first from *Aspergillus ochraceus* in 1965 (Van der Merwe *et al.*, 1965). It has also been demonstrated that this molecule could be produced by other *Aspergillus* species such as *A. carbonarius* (Belli *et al.*, 2005), *A. alliaceus* (Bayman *et al.*, 2002) and *A. niger* (Abarca *et al.*, 1994), although the frequency of toxigenic strains in this species appears to be moderate (Teren *et al.*, 1996). OTA can also be synthesized by *Penicillium* species, mainly *P. verrucosum* (previously named *P. virridicatum*) (Pitt, 1987). The ability of both *Aspergillus* and *Penicillium* species to produce OTA makes it a worldwide contaminant of numerous foodstuffs. Indeed, *Aspergillus* is usually found in tropical or subtropical regions, whereas *Penicillium* is very common contaminant in temperate and cold climate areas (Pardo *et al.*, 2006). Many surveys have revealed the contamination of large variety of vegetal products; such as cereals (Jorgensen, 2005), grapes (Battilani *et al.*, 2006) and coffee (Taniwaki, 2006). For cereals, OTA contamination generally occurs during storage of raw materials, especially when moisture and temperature are abnormally high (Magan and Aldred, 2005).

2.2.2 Impact on poultry performance

Acute forms of intoxication have been reported after exposure to concentrations of OTA ranging from 2 to 10 mg/kg feed. A decrease in feed intake has been noted in turkeys but not in chickens (Hamilton *et al.*, 1982; Burditt *et al.*, 1984). The animals were prostrate, ataxic with muscular trembling and displayed reduced reflexes. Mortality reached 55% and post mortem dissection revealed that kidneys were pale, swollen and haemorrhagic (Hamilton *et al.*, 1982).

Chronic forms of toxicity are observed after exposure to OTA at concentrations ranging from 0.3 to 4 mg/kg feed for a period of several weeks. Such concentrations are quite high, the minimal toxic levels for poultry is considered to be close to 0.5 mg/kg feed in both laying hens and broilers (Huff *et al.*, 1974). Administration of 0.2 ppm of OTA for 20 days did not result in the appearance of macroscopic or microscopic lesions, nor to persistence of residues (LD = 0.0005 mg/kg) in broilers (Kozaczynski, 1994). The main symptoms observed are retarded growth and a reduced feed conversion rate. Nephropathy has been reported in all species after exposure to concentrations equal to or higher than, 2 mg OTA/kg feed (Chang *et al.*, 1981; Hamilton *et al.*, 1982; Dwivedi and Burns, 1984; Kubena *et al.*, 1985). This is accompanied by an increased water intake, the clinical sign of which is the emission of large quantities of moist faeces (Elling, 1975). Digestive symptoms such as an increase in gizzard weight and deficient muscular conformation with a delay in sexual maturity have also occasionally been reported (Huff *et al.,* 1988). In chicken, a loss of pigmentation, probably due to a decrease in plasmatic concentration of carotenoids, and a decline in bone strength, as well as osteoporosis, have also been observed (Hamilton *et al.*, 1982; Dwidedi and Burns, 1984; Duff *et al.*, 1987). In laying hens, a reduction in egg production has been observed without feed refusal (Prior and Sisodia, 1978). Eggs were also seen to have an abnormally high number of spots; with shells which were thin and fragile (Page *et al.*, 1980; Hollinger and Ekperigin, 1999). A higher protein content or a supplementation of the diet with phenylalanine reduced these clinical signs (Gibson *et al.*, 1989; Gibson *et al.,* 1990).

These effects are accompanied by several biochemical and haematological changes. A decrease in serum content of proteins, cholesterol, triglycerides, ammonia, calcium, inorganic phosphates, potassium and carotenoids, as well as an increase in the concentration of uric acid and creatinine linked with an increase in glomerular filtration have been reported (Huff and Hamilton, 1975; Page *et al.*, 1980; Schaeffer *et al.*, 1987; Huff *et al.,* 1988; Glahn *et al.*, 1988; Elissalde *et al.*, 1994). A hypochromic-microcytic anemia has also been reported Huff *et al.*, 1988) as well as an alteration of blood coagulation (Doerr *et al.*, 1981). Once again, an increase in protein content or supplementation of the diet with phenylalanine, decreased biochemical and haematological signs of ochratoxin A toxicity (Bailey *et al.*, 1990).

2.2.3 Impact on immunity

A reduction in immune defences has been noted in poultry exposed to ochratoxin A. It can be characterized by leucocytopenia, medullar hypoplasia and a decrease in the leukocyte count of immune tissues (bursa of Fabricius, thymus, spleen, etc.) (Dwivedi and Burns, 1984a). There is also a reduction in the concentration of immunoglobulins; and in the phagocytic capacity of leucocytes and neutrophilic polynuclear cells (Dwivedi and Burns, 1984b). An increase in sensitivity to bacterial and parasitic infections has also been reported. For example, exposure of chickens to OTA increased sensitivity to both *Salmonella* and *E. coli* infection (Fukata *et al.*, 1996; Kumar *et al.*, 2003, 2004). OTA also increases the severity of the effect *Eimeria* has on the body weight of infected broiler chickens (Huff and Ruff, 1982). However, it has to be noted that all these studies were done with high levels of toxins, usually a few ppm and little is known about the impact of OTA on the immune system at concentrations found in natural contamination of feeds (several ppb).

2.3. Fumonisins

Since 1973, several poultry diseases have been associated with the consumption of maize contaminated with *Fusarium moniliforme* or *F. proliferatum* fungi. A decline in zootechnical performance, feed refusal, ataxia, diarrhoea and even mortality, has been observed (Sharby *et al.*, 1973). Cases of mortality after ingestion of grains contaminated with *F. moniliforme* have also been reported in wild birds (Weibking *et al.*, 1993).

2.3.1 Origin

Fumonisins were first described and characterized in 1988 from *Fusarium verticillioides* (formerly *F. moniliforme*) culture material (Bezuidenhout *et al.*, 1988; Gelderblom *et al.*, 1988). The most abundant and toxic member of the family is fumonisin B$_1$ (FB$_1$). These molecules can be produced by only a few species of *Fusarium* fungi, namely *F. verticillioides, F. proliferatum* and *F. nygamai* (Marin *et al.*, 2004). These fungal species are worldwide contaminants of maize that represent the main source of fumonisins in feed (CSHFP, 1999). *Fusarium verticillioides* grows as an endophyte on corn and can causes plant pathology such as seedling blight, stalk rot or ear rot. However, fumonisin contamination of the grains themselves, can occur without any visible signs. Fumonisin production mainly occurs during the pre-harvest period, at temperatures of 20-25 °C and when the moisture content of grains is high (Le Bars *et al.*, 1994; Marin *et al.*, 2004).

2.3.2 Impact on poultry performance

As is true for other animal species, the effects of fumonisins in poultry are related to the dose and the length of exposure. In the short term, administration of high concentrations of fumonisin B_1 (>100 mg/kg feed) can increase mortality, with a more serious effect in young animals. Normally, ingestion of feed contaminated with 10 to 300 mg/kg feed only leads to a decrease in feed conversion rate, average body weight (Figure 1), and some modification of organ weight. Blackish sticky diarrhoea has been reported and could be related to the decrease in feed digestibility (Table 2).

A recent study on the impact of chronic exposure to fumonisins in turkeys demonstrated that the exposure of this species to 20 mg FB_1/kg feed, corresponding to the highest dose allowed by EU regulation, did not lead to detectable modification of body weight or feed consumption (Tardieu *et al.*, 2007). By contrast, ducks appear to be more sensitive to fumonisins since the exposure to comparable concentrations of toxins (32 ppm), led to a significant reduction in growth (Tran *et al.*, 2005). Interestingly, this effect appeared to be more pronounced when the toxin was administered over 1 to 7 weeks, than when it was ingested for more than 10 weeks (Tran *et al.*, 2005). These results suggest a certain degree of 'adaptation' to fumonisins by this animal species. A study performed during forced feeding of mule ducks, also revealed that the ingestion of maize containing 20 ppm FB_1, significantly reduced breeding performances and led to serious economic losses for producers (Tardieu *et al.*, 2004).

Figure 1. Effect of fumonisin B_1 on body weight of mule ducks receiving 0 (left) and 128 mg FB_1/kg feed (right). (A) after 14 days of exposure and (B) after 35 days of exposure (Tran et al., 2005).

Table 2. Effects of fumonisin B$_1$ in poultry.

Dose	Duration of exposure	Species, age	Symptoms, lesions (higher doses)	References
5, 15, 45 mg/kg BW	12 days	28-day-old duck	Decrease in body weight gain, increase in liver weight with appearance of acino-tubular structures. Induction of several enzymatic pathways.	Bailly et al., 2001 Raynal et al., 2001
100 to 500 mg/kg feed	14 to 420 days	1-day-old chick 1-day-old duckling 1-day old turkey Laying hens	Decrease in body weight and mean daily weight gain. Increase in relative weight of liver, kidney, spleen and proventriculus, Hepatocytes proliferation with hyperplasia. Diarrhoea. Decrease in antibodies titre.	Brown et al., 1992 Ledoux et al., 1992 Weibking et al., 1993 Bermudez et al., 1995 Kubena et al., 1995, 1997, 1999 Ledoux et al., 1996 Li et al., 2000
2-50 mg/kg feed	6 days to 11 weeks	1-day-old chick Chick Turkey Ducks	Sticky darkish diarrhoea, decrease in body weight and daily feed intake. Increase in relative weight of liver and spleen; reduction in fat content of liver. Decrease in relative weight of proventriculus. No histological lesion of liver or kidney No modifications of breeding performances	Prtathapkumar et al., 1997 Espada et al., 1994 Henry et al., 2000 Broomhead et al., 2002 Tran et al., 2005 Tardieu et al., 2004, 2007

Several recent studies were performed to evaluate, in breeding conditions, the risk of persistence, at the residual level, of fumonisins in edible parts of poultry. From these experiments, it appears that the concentration of 20 mg fumonisins/kg of feed, that is recommended as maximal value in poultry feed by E.U. regulation, is associated with a residual contamination of livers with fumonisins (Table 3) (Tardieu *et al.*, 2008; Galtier *et al.*, 2008). This is in agreement with the reported contamination of avian liver with fumonisins, revealed in a recent global survey on food safety, in France (Leblanc *et al.*, 2005).

Besides the effects on growth and performance, several biochemical modifications have also been reported. For example, in growing ducks, a transient increase in aspartate aminotransferase (AST), alanine aminotransferase (ALT), lactate dehydrogenase (LDH), proteins and cholesterol, were described after exposure to fumonisins. This was followed by a progressive return to normal values, in the case of prolonged exposure (Tran *et al.*, 2005). The most important biochemical modification following fumonisin exposure is the increase of sphinganine to sphingosine ratio (Sa:So). Indeed, in all studies where it has been investigated, this appears to be the most sensitive and earliest biomarker, of exposure to fumonisin. It allows the detection of animals exposed to weak concentrations of the toxin, before the appearance of any clinical symptoms or impact on animal performance (Tran *et al.*, 2003, 2006; Tardieu *et al.*, 2006).

Table 3. Residual persistence of fumonisin B_1 (µg/kg) in turkey and ducks (Galtier et al., 2008).

Tissues	Fumonisins concentration in feeds (mg/kg) and tissular content (µg/kg)			
	0	5	10	20
Turkeys				
Liver	<LD[1]	33 ± 30	44 ± 20	117 ± 50
Kidney	<LD	<LD	<LD	22 ± 8
Muscles	<LD	<LD	<LD	<LD
Ducks				
Liver	<LD	<LD	16 ± 3	20 ± 6
Kidney	<LD	<LD	<LD	<LD
Muscles	<LD	<LD	<LD	<LD

[1] Limit of detection (LD) = 12,5 µg/kg.

2.3.3 Impact on immunity

Various alterations of the immune system have also been reported *in vivo* and *in vitro*; decrease in the thickness of the thymal cortex, decrease in immunoglobulin content (Qureshi *et al.*, 1995), decrease in peripheral lymphocytes viability (Dombrink-Kurtzman *et al.*, 1993), modification of morphology and functionality of macrophages (Qureshi and Hagler, 1992), an alteration in bacteria elimination capacity and a decrease in response to vaccines (Javed *et al.*, 1995; Li *et al.*, 2000). Once again, all these studies were performed using high doses of fumonisins compared to that observed in natural contamination cases.

2.4. Trichothecenes

As is true for most farm species, poultry intoxication by trichothecenes can appear in an acute or chronic form. Although acute toxicity is quite easy to diagnose, it is also the rarest form. Diagnosis of chronic intoxication is more difficult but this is the form which is mainly found in conditions of natural contamination by trichothecenes. Trichothecenes include deoxynivalenol (DON), T-2 toxin, diacetoxyscripenol (DAS) and HT-2 toxin.

2.4.1 Origin

Trichothecenes can be produced by a large variety of fungi. They mainly belong to the *Fusarium* genus, however other fungal species such as *Trichoderma viridae* and *Myrothecium roridum* have also been shown to be able to produce some trichothecenes (Bean *et al.*, 1992; Wilkins *et al.*, 2003). In addition, one fungal species may be able to produce several types of trichothecenes. These fungal species mainly belong to the field mycoflora, developing on living plants or during the early post-harvest period. Indeed, *Fusarium* species are hygrophilic fungus and drying will stop their development (Schrödter, 2004). These fungal contaminants are well known for being responsible for *Fusarium* head blight of small grain cereals and ear rot of maize. These fungal pathologies reduce yields, decrease milling and malting qualities of grains, and may lead to mycotoxin contamination of the infected grains (Parry *et al.*, 1995; Logrieco *et al.*, 2002).

2.4.2 Impact on poultry performance

The main symptoms observed in the case of acute intoxications were nervous and digestive disorders. Within minutes of administration hyperpnoea occurred, along with lethargy and loss of balance. The main lesions were a haemorrhagic syndrome in the digestive tract and muscles (Huff *et al.*, 1986; Richardson and Hamilton, 1990). Symptoms linked to chronic trichothecene intoxication are not indicative. The nervous disorders and diarrhoea observed in cases of acute intoxication are not described. Intoxication has only been reported to lead to a reduction in performance, retarded growth and a decrease in egg production. The effects of DON and T-2 toxin are reported in Tables 4 and 5, as a function of the dose and length of exposure.

In growing animals, it is interesting to note that there are some discrepancies between experimental results, mainly in the effect on mean growth rate and feed conversion ratio. Nevertheless, analysis of the results usually shows a decrease in the mean growth rate and, contrary to what is reported in pigs, an increase in feed conversion. In laying hens, the main effects observed were partial feed refusal with no effect on the final weight but this was accompanied by a decrease in the laying rate and the production of abnormally fragile eggs. The decrease in egg production was also accompanied by a reduction in hatchability and web-footed birds seem to be more sensitive than gallinacean.

Table 4. Effect of chronic exposure to DON in poultry.

Species, dose (mg/kg feed), duration of exposure	Signs and/or lesions	References
Broiler, 0.3-1.87 mg/kg, 4 weeks	No sign.	Hulan *et al.*, 1982
Laying hen, 0.12-4.9 mg/kg, 10-24 weeks	No sign (only smaller and fragile eggs with 0.7 mg/kg for 10 weeks). Increase in malformation rate at hatching.	Bersjo *et al.*, 1993 Hamilton *et al.*, 1985
Broiler, 16 mg/kg, 3 weeks	Decrease in the daily mean weight gain, increase in the feed conversion ratio. Necrosis of oral cavity. Increase in relative gizzard weight and on weight of bursa of Fabricius. Decrease in red blood cell numbers and haematocrit. Decrease in mean globular volume. Decrease in triglycerides and LDH activity.	Huff *et al.*, 1986 Kubena *et al.*, 1989

Table 5. Effect of prolonged exposure to T-2 toxin in poultry.

Species, dose and duration of exposure	Signs and/or lesions	References
Broiler, 0.2-4 mg/kg, 3-9 weeks	No clinical signs, transient increase in daily body weight gain at 0.2 mg/kg during the first 6 weeks. No modification of feed conversion ratio.	Chi *et al.*, 1977
Duck, 0.2-4 mg/kg, 7 weeks	Necrosis of tongue, palate, mouth and pharynx at highest doses. 50% mortality at 4 mg/kg. Decline in lymphocyte response to mitogenic agents, lymphocytic depression in lymphoid organs (for animals treated with 3-4 mg/kg). Decrease in daily body weight gain and in feed intake.	Rafai *et al.*, 2000
Goose, 0.2-3 mg/kg BW, 18 days	Decrease in egg production and hatchability. 10-70% mortality.	Vanyi *et al.*, 1994
Laying hen, 0.5-10 mg/kg, 8 weeks	Decrease in feed intake. Decrease in egg production, increase in number of infertile eggs (1st week) and decrease in hatchability (from week 3) depending on the concentration. Necrosis of oral cavity. Ulcer in anterior part of gizzard (4-8 mg/kg).	Chi *et al.*, 1978 Diaz *et al.*, 1994
Turkey, 3-10 mg/kg, 4 weeks	Decrease in daily body weight gain. Necrosis of oral cavity. Decrease in the size of Fabricius bursa, accelerated thymus involution.	Richard *et al.*, 1978
Broiler, 2-16 mg/kg, 2-4 weeks	Decrease in daily body weight gain and in feed intake. Increase in feed conversion ratio. Increase in the weight of Fabricius bursa. Necrosis of oral cavity from third week.	Huff *et al.*, 1988 Kubena *et al.*, 1989 Wyatt *et al.*, 1973 Richard *et al.*, 1978

Lesions observed in the case of chronic intoxication by trichothecenes are suggestive enough to lead to a diagnosis, when a decrease in productivity has been seen on farm. The severity of lesions and the speed of their appearance, as well as the minimal dose of toxin needed, depends not only on the necrotic capacity of the molecules but also on the species of animal. The length of the period before the appearance of symptoms can range from few days, in very sensitive species such as duckling or in the case of high doses, to several weeks. Sometimes, necrotic lesions are only perceptible after histological examination of organs. Only a few studies have reported biochemical and haematological alterations. The only alterations that are regularly observed appear to be a decrease in LDH and alkaline phosphatase (ALP) activity, as well as a decrease in protein and albumin, probably as a result of the modification of protein synthesis.

2.4.3 Impact on immunity

Although the impact of trichothecenes on immune function have been extensively studied in laboratory animals, only few direct measurements of their effects on immune function in birds have been carried out. They are reported in Tables 4 and 5. Depending on the dose, toxin and animal species, different results have been reported. For example, exposure of turkeys or broilers to 8-16 mg T-2 toxin/kg of feed decreased the size of the bursa of Fabricius (Wyatt *et al.*, 1973; Richard *et al.*, 1978). By contrast, in broilers, exposure to 4 mg T-2 toxin/kg of feed and 16 mg DON/kg of feed increased the size of this organ (Kubena *et al.*, 1989b). T-2 toxin also decreases the lymphocytes response to mitogenic agents (Rafai *et al.*, 2000).

The immunosuppressive effects of the trichothecenes on cellular and humoral mediated immune response have been shown to decrease host resistance to infectious diseases (Corrier, 1991). In chickens, mortality caused by salmonellosis was reported to increase in cases of T-2 toxin exposure (Boonchuvit *et al.*, 1975). However, more recently, a study carried out with lower concentrations of T-2 toxin (<1 ppm) for 28 days, did not reveal any impairment of immune competence as measured by antibody production (Sklan *et al.*, 2003).

2.5. Zearalenone

Among all species that have been studied, poultry appear to be the most resistant to zearalenone (ZEA) (Gaumy *et al.*, 2001).

2.5.1 Origin

Zearalenone was isolated for the first time from maize contaminated with *Gibberella zeae*, the anamorph of *Fusarium graminearum* (Stob *et al.*, 1962). It has been demonstrated that it can be synthesized by several *Fusarium* species including *F. graminearum*, *F. proliferatum*, *F. culmorum* and *F. oxysporum* (Sydenham *et al.*, 1991; Molto *et al.*, 1997). As observed for fumonisins and trichothecenes, these fungal species usually develop on living plants and ZEA contamination occurs in the field, at harvest or early storage, when drying has not been adequate. Temperature of ZEA production is lower than optimal temperature for mycelium development, about 20-25 °C (Llorens, 2004). ZEA production is favoured in substrates with high glucid to protein ratio.

2.5.2 Impact on poultry performance and immunity

During experimental exposure, concentrations higher than 100 mg/kg of feed are usually required to induce clinical signs, these are summarized in Table 6.

Turkeys appear to be the most sensitive avian species. A recent study on zearalenone metabolism in different avian species revealed significant differences

Table 6. Effects of zearalenone in poultry.

Zearalenone	Duration of exposure	Species, sex and age	Symptoms, lesions
1-10-30 mg/kg	4 weeks	Chickens/quails	No symptom, no lesion
10-25 mg/kg	3 weeks	Turkeys, 3 weeks old	No symptom, no lesion
25-100 mg/kg	28 days	Breeding hens, 20 weeks old	No effect on reproductive function
50-200-400-800 mg/kg	7 consecutive days	Hens	Linear increase in oviduct weight
50-100-400-800 mg/kg	3 weeks	Broilers, 6 weeks old	No lesion
10-25-50-100-200-400-800 mg/kg	chronic	Laying hens, 30 weeks old	No effect on reproductive function
300 mg/kg	4 days	Turkeys, 10 days old	Increase in weight gain
400-800 mg/kg	chronic	Adult male turkeys	Development of wattles, no biochemical nor haematological signs
100-800 mg/kg	chronic	Adult cocks	Normal semen, decrease in phosphate, cholesterol and AST

in toxin metabolisation that may explain the differences in sensitivity to this mycotoxin (Kolf-Clauw *et al.*, 2008). In young animals, an anabolic effect was noted in 10 to 12-day-old animals fed with a diet containing 300 mg/kg of zearalenone. An increase in mucosal secretion in faeces, hyperplasia of the oviduct and the cloaca, followed by an reversion of the latter after 4 days of treatment, were also observed. In adults, administration of 100 mg/kg zearalenone for 56 days led to a 20% decrease in egg production in females (Allen *et al.*, 1983), while the administration of 400 or 800 mg/kg of feed, increased dewlaps and caruncles in males (Allen *et al.*, 1981). The economic impact of zearalenone in industrial breeding conditions has not been documented. Moreover, even if alteration of immunological parameters were observed *in vitro*, no impact on animal immunity of this mycotoxin has been demonstrated until now.

2.6. Other toxins of interest in poultry

2.6.1 Moniliformin

Moniliformin is a toxin produced by *Fusarium* species such as *F. verticillioides*, *F. subglutinans* and *F. proliferatum* (Hussein *et al.*, 1991; Fotso *et al.*, 2002). The oral LD 50 for this toxin has been found to be 5.4 mg/kg of body weight (Burtmeister *et al.*, 1979). Therefore, acute toxicity of this molecule is comparable to that of T-2 toxin but chronic toxicity is lower. For instance, a decrease in body weight gain was reported for dietary doses of 64 mg/kg of feed and 100 mg/kg of feed, for female and male broilers respectively (Allen *et al.*, 1981; Harvey *et al.*, 1997; Li *et al.*, 2000). The similar concentrations were required to observe adverse effects on performance in other poultry species (Kubena *et al.*, 1999). Classical lesions consist of an increased heart weight with, at the histological level, loss of cardiomyocyte striations (Harvey *et al.*, 1997; Bermudez *et al.*, 1997; Ledoux *et al.*, 2003). Dietary administration of moniliformin at 100 mg/kg of feed led to a marked reduction in egg production in laying hens (50%) and a transient reduction in egg weight (Kubena *et al.*, 1999). This concentration also altered immune function in broiler chicks, for example, it led to a decrease in vaccine reaction. This occurred along with a reduction in antibody titres and an increase in the sensitivity of broiler chicks to infection, possibly by reducing the proliferation of lymphocytes, as demonstrated in vitro (Li *et al.*, 2000).

2.6.2 Cyclopiazonic acid

Cylopiazonic acid was first isolated from a culture of *Penicillium cyclopium* but has also been shown to be produced by several species of *Aspergillus* and

Penicillium such as *A. flavus*, *A. tamarii* and *P. camemberti* (Le Bars, 1979; Martins and Martins, 1999). Even though cylopiazonic acid has been found in many feeds and foods, only few cases of intoxigation have been described.

The LD 50, following oral administration, is 20 mg/kg BW in broiler chicks (Vesely *et al.*, 1985). Acute effects of cyclopiazonic acid seem to be reversible if the ingested dose is not too high. The main symptoms are nervous disorders such as ataxia, apathy, muscular spasms. In 3-week-old broilers, administration of 50 mg cyclopiazonic acid/kg of feed for 3 weeks led to a significant decrease in body weight gain from the first week of exposure. This reduction was progressive, suggesting a cumulative toxicity of the molecule. A reduction in the relative weight of the bursa of Fabricius has also been reported, together with an increase in the weight of the liver, kidney and proventriculus. Significant increases in uric acid and cholesterol concentrations have been observed (Smith *et al.*, 1992). The same observations were made in 1 to 3-week old broilers after exposure to 34 mg/kg of feed for 3 weeks (Gentles *et al.*, 1999).

Moreover, retrospective analysis of 'Turkey X disease' performed in 1986 by Cole, suggested that clinical signs were not all typical of aflatoxicosis. He therefore tried to demonstrate a possible role for cyclopiazonic acid in this intoxication. For instance, opisthotonos originally described in 'Turkey X disease' can be reproduced by administration of a high dose of cyclopiazonic acid but not by ingestion of aflatoxin (Cole, 1986). Post-mortem examination of birds usually revealed thickened mucosa and dilated proventricular lumen (Smith *et al.*, 1992; Gentles *et al.*, 1999). At a concentration of 100 mg/kg feed, ulcerative proventriculitis, mucosal necrosis of the gizzard, hepatic and splenic necrosis were observed after 7 weeks of exposure (Dorner *et al.*, 1983).

2.6.3 Combination of toxins

Since poultry feeds are made of several different raw materials, exposure to a mixture of mycotoxins can occur in field conditions. Several studies investigated the combined effects of mycotoxins in the case of co-contamination of the feed. Most often, these studies were performed with high levels of contamination and did not aim to describe what could be observed in circumstances of natural contamination. Available data indicates a synergistic effect of OTA and AFB_1 (Warren and Hamilton, 1980; Campbell *et al.*, 1983; Huff *et al.*, 1984; Verma *et al.*, 2004), and of T-2 toxin and AFB_1 (Huff *et al.*, 1988a). The effects are said to be additive for OTA and trichothecenes (DON, T-2 toxin, diacetoxyscirpenol) (Kubena *et al.*, 1989; Kubena *et al.*, 1994), AFB_1 and FB_1 (Tessari *et al.*, 2006),

FB$_1$ and DON (Kubena *et al.*, 1997), and for OTA and cyclopiazonic acid (Gentles *et al.*, 1999).

Recent studies demonstrated that a combination of toxins, at naturally occurring concentrations, could lead also lead to a reduction in animal performance and especially body weight gain (Danicke *et al.*, 2007; Girish *et al.*, 2008b). It can also lead to more subtle effects in exposed animals such as modification of intestinal morphology during the growth phase (Girish and Smith, 2008), modification of some immunological parameters (Girish *et al.*, 2008b) and alteration of pons serotonergic system (Girish *et al.*, 2008a). All avian species do not display same sensitivity, ducklings appearing more resistant than turkeys or broilers to a mixture of trichothecenes and zearalenone (Chowdhury *et al.*, 2005ab; Yegani *et al.*, 2006).

Such studies demonstrate that, in farming conditions, deleterious effects can be observed at lower doses that those reported to have an effect when using pure toxins. However, the results may also be difficult to interpret since the reported effects may be related to the presence of toxins other that those which were quantified. Consequences of mould growth such as a reduction in the digestibility or nutritional value of grains, can also affect animal performance. The characterisation of the possible impact of the exposure to a mixture of toxic molecules at naturally occurring concentration is of major interest to develop efficient prevention strategies for producers.

3. Natural occurrence of mycotoxins in poultry feeds and strategies for prevention

The risk of mycotoxicosis in poultry production is due to natural contamination of the feed by mycotoxins. These compounds can be present in raw materials before the diet is formulated or can be synthesised during storage of feeds. Mycotoxins are synthesised during fungal growth and sporulation. Taking into account their eco-physiological characteristic, fungal species are usually classified in 'fungal flora of the field' (*Fusarium* species and, *Aspergillii* in hot and wet climate countries) or 'storage flora' (*Penicillia* and *Aspergillii*) (Wilson *et al.*, 2002; Pitt, 2002; Miller, 2002). Therefore, contamination of poultry feeds by mycotoxins has to be investigated at two separate levels; contamination of raw materials in the field and mycotoxin synthesis during storage, before consumption by animals.

3.1. Contamination of raw materials

Poultry feeds are mainly composed of cereals with a high proportion of wheat (20 to 30%), maize (10 to 20%) and barley (about 15%). All of these plants are very susceptible to *Fusarium* infection and *Fusarium* toxin contamination.

A recent comprehensive survey on *Fusarium* toxin contamination of cereals in Europe, was carried out to assess dietary intake of the population of EU member states and revealed a high prevalence of mycotoxin contamination in the raw materials throughout Europe (Table 7) (Gareis *et al.*, 2003). Although the levels of contamination observed are generally low, they raise the question of possible chronic toxicity in animals due to constant exposure to these important components of the diet (45 to 60% of poultry feed intake). Deoxynivalenol and fumonisins appear to be of major concern in European poultry production, due to their impact on zootechnical performance of birds in the case of chronic exposure.

Surveys done in other parts of the world, mainly in countries with warmer climates, also revealed that cereals may be contaminated by aflatoxins (Nizamlyoglu and Oguz, 2003), sometimes with a high prevalence (40 to 80% of samples) and high levels of contamination (above national or European regulatory limits) (Zinedine *et al.*, 2007). However, following extreme climatic conditions, such as above average temperatures in summer, aflatoxins could

Table 7. Contamination of cereals by Fusarium toxins in Europe (Gareis et al., 2003).

Fusarium toxin	Main food contaminated (% of positive samples)
Type B trichothecenes	
Deoxynivalenol	Maize (89%), wheat (61%)
Nivalenol	Maize (35%), oats (21%), wheat (14%)
3-acetyldeoxynivalenol	Maize (27%), wheat (8%)
Type A trichothecenes	
T-2 toxin	Maize (28%), wheat (21%), oats (21%)
HT-2 toxin	Oats (41%), corn (24%), rye (17%)
Zearalenone	Maize (79%), wheat (30%)
Fumonisins	
Fumonisin B_1	Maize (66%)
Fumonisin B_2	Maize (51%)

be found in other part of the world. For example, in 2003, controls on maize harvested in Europe were found contaminated by unusual levels of AFB1 (Battilani *et al.*, 2005; Giorni *et al.*, 2007) whereas European crops are usually considered aflatoxin free.

3.2. Contamination of poultry feeds

Surveys of fungal flora and mycotoxin contamination of poultry feeds can highlight contamination that may occur in the field but also the quality of storage procedures. Indeed, the predominant genus isolated is *Penicillium*, followed by *Aspergillus*, which can be considered as part of the storage flora (Oliveira *et al.*, 2006). *Fusarium* spp, which are frequently found in raw materials, are less present in feeds because the granulation step usually destroys fungal spores before mycotoxins are observed (Labuda *et al.*, 2005; Labuda *et al.*, 2006; Fraga *et al.*, 2007). Available data shows that a relatively low level of contamination of feeds with mycotoxins, would be unable to lead to clinical signs in poultry or to significant contamination of edible parts of the animals (Cespesdes *et al.*, 1997; Beg *et al.*, 2006). However, some studies have reported levels of contamination compatible with the appearance of chronic or sub-clinical toxicity of; aflatoxin (several tens of µg/kg of feed), deoxynivalenol (several mg/kg of feed) or fumonisin B_1 (tens of mg/kg feed), in poultry species (Nizamlyoglu and Oguz, 2003; Tardieu *et al.*, 2004; Oliveira *et al.*, 2006; Zinedine *et al.*, 2007). Moreover, the possible impact of a combination of toxins on immune function and or breeding performances remains to be clarified.

3.3. Prevention strategies

Many strategies have been developed to prevent the detrimental effects of mycotoxins in farm animals. These methods aim to either:
• prevent mycotoxin synthesis and accumulation;
• detoxify the toxins present in feeds;
• inhibit the absorption of ingested toxins in the gastrointestinal tract of animals.

3.3.1 Prevention of mycotoxin contamination

Mycotoxin contamination may occur in the field, during the pre-harvest period or during storage of raw materials and feeds, depending of eco-physiologic properties of the producing fungal species. At the pre-harvest level, the major hazard for cereals is the development of toxigenic *Fusarium* species which are able to produce trichothecenes, fumonisins or zearalenone. Aflatoxins can also contaminate cereals at the field level. Climatic conditions are a key factor that directly influences contamination level of cereals with mycotoxins (Cotty and Jaime-Garcia, 2007). However, several strategies have been developed to limit mould development and toxinogenesis.

Many studies have aimed to find and select cereal varieties that are resistant to mould infection (Bai *et al.*, 2001; Wu, 2006). Mycotoxin control is also based on the improvement of agricultural practices and field management (Aldred and Magan, 2004). For example, fungal development and subsequent thrichothecene production may be related to agricultural practices such as crop rotation (Edwards, 2004). It is generally accepted that wheat that follows an alternative host for *Fusarium* pathogen (i.e. maize) is at greater risk for DON contamination of grain (Obst *et al.*, 2000). Therefore, European commission recently published recommendations on agricultural practices that hope to limit the contamination of crops with *Fusarium* toxins (EU, 2006).

Post-harvest measures are also important to limit mycotoxin contamination of raw material and feeds during the storage period. Moisture of grains and temperature appear to be the most critical factors in the growth of mycotoxinogenic molds (Magan *et al.*, 2003).Therefore, storage facilities have to protect crops from rain, provide drainage of ground water and avoid entry of rodents, birds and insects, as well as limiting temperature fluctuations (Codex Alimentarius, 2002). Various natural and chemical agents may also be used to limit mould growth and toxinogenesis during storage. More recently, biological competition with atoxigenic moulds, bacteria or yeast has received a lot of attention and should be a promising approach as an alternative to fungicide treatment (Kabak *et al.*, 2006)

3.3.2 Detoxification of mycotoxins

Numerous strategies based on physical or chemical treatment of raw materials or feeds have been developed to decrease mycotoxin content of contaminated feeds. However, most mycotoxins are quite stable and physical treatments within the range of conventional food processing temperatures are of limited efficacy (Scott, 1998). At least for aflatoxins, chemical detoxification based on an ammoniation procedure was shown to be effective (Piva *et al.*, 1995). More recently, ozone treatment has been developed and appears to be able to degrade several mycotoxins including trichothecenes and aflatoxins (Proctor *et al.*, 2004; Young *et al.*, 2006).

3.3.3 Adsorption of mycotoxins

One of the most recent approaches to control deleterious effects of mycotoxins in livestock is to limit mycotoxin absorption in the intestinal tract. This is done by the addition of molecules able to bind mycotoxins, to the diet and subsequently reduce their bioavailability. The major interest in this approach is that it may allow the control of several toxins with only one molecule. It may also be of particular interest in countries where climatic conditions and cereal production process are favourable for mycotoxin synthesis.

Activated carbons, Hydrated sodium calcium aluminosilicate, zeolites and bentonites have all been extensively studied for their affinity for mycotoxins (Guerre *et al.*, 2000; Huwig *et al.*, 2001; Kabak *et al.*, 2006). Recently, a yeast cell wall derived glucomannan was tested for its ability to prevent toxicity of grains naturally contaminated with a mixture of different mycotoxins. The addition of 0.2% of this adsorbant was shown to prevent most of the deleterious effects induced by contaminated grains in various avian species (Yegani *et al.*, 2006; Girish *et al.*, 2008b; Girish and Smith, 2008).

4. Conclusion

Although poultry are usually considered to be relatively resistant to mycotoxins, ingestion of high concentrations of mycotoxins can have many detrimental effects. However, natural contamination of raw materials and or poultry feed is usually not high enough to cause acute toxicity.

Nevertheless, prolonged exposure to low concentrations of mycotoxins can induce chronic toxicity which is mostly characterized by a decrease in feed consumption, mean daily weight gain, feed conversion ratio and egg production. Another problem is the potential impact of mycotoxins on the immune functions of poultry, which can for example, increase susceptibility to infection or reduce the efficiency of vaccination. The clinical signs are not always very clear, but they can nevertheless rapidly result in serious financial losses for producers.

It appears that in European climatic conditions, deoxynivalenol and fumonisins are the main cause of concern in poultry feed. However, other mycotoxins, such as aflatoxins, occur worldwide and could lead to chronic or sub-clinical toxicity during production. Moreover, possible interactions between different mycotoxins, particularly at low concentrations, remain to be elucidated. In order to evaluate the real consequences of mycotoxin contamination of poultry feed on avian health, immunity and performances, these interactions need to be studied and adequate prevention strategies developed.

References

Abarca, M.L., Bragulat, M.R., Castella, G. and Cabanes, F.J. (1994). Ochratoxin A production by strains of *Aspergillus niger* var *niger*. *Applied and Environmental Microbiology* **60**: 2650-2652.

Aldred, D. and Magan, N. (2004). Prevention strategies for trichothecenes. *Toxicology Letters* **153**: 165-171.

Allen, N.K., Mirocha, C.J., Aakhus-Allen, S., Bitgood, J.J., Weaver, G. and Bates, F. (1981). Effect of dietary zearalenone on reproduction of chickens. *Poultry Science* **60**: 1165-1174.

Allen, N.K., Peguri, A., Mirocha, C.J. and Newman, J.A. (1983). Effects of *Fusarium* culture, T-2 toxin and zearalenone on reproduction of turkey females. *Poultry Science* **62**: 282-289.

Bai, G.H., Plattner, R., Desjardins, A. and Kolb, F. (2001). Resistance to *Fusarium* headblight and deoxynivalenol accumulation in wheat. *Plant Breeding* **120**: 1-6.

Bailey, C.A., Gibson, R.M., Kubena, L.F., Huff, W.E. and Harvey, R.B. (1990). Impact of L-phenylalanine supplementation on the performance of three week old broiler fed diet containing ochratoxin A. Effect on haematology and clinical chemistry. *Poultry Science* **69**: 420-425.

Bailly, J.D., Benard, G., Jouglar, J.Y., Durand, S. and Guerre, P. (2001). Toxicity of *Fusarium moniliforme* culture material containing known levels of fumonisin B1 in ducks. *Toxicology* **163**: 11-22.

Battilani, P., Scandolara, A., Barbano, C., Pietri, A., Bertuzzi, T., Marocco, A., Berardo, N., Vannozzi, G.P., Baldini, M., Miele, S., Salera, E. and Maggiore, T. (2005). Monitoraggio della contaminazione da micotossine in mais. *Infor. Agro.* **61**: 47-49.

Battilani, P., Magan, N. and Logrieco, A. (2006). European research on ochratoxin A in grapes and wine. *International Journal of Food Microbiology* **111**: S2-S4.

Bayman, P., Baker, J.L., Doster, M.A., Michailides, T.J. and Mahoney, N.E. (2002). Ochratoxin production by the *Aspergillus ochraceus* group and *Aspergillus alliaceus*. *Applied and Environmental Microbiology* **68**: 2326-2329.

Bean, G.A., Jarvis, B.B. and Aboul-Nasr, M.B. (1992). A biological assay for the detection of *Myrothecium* spp. produced macrocyclic trichothecenes. *Mycopathologia* **119**: 175-180.

Beg, M.U., Al Mutairi, M., Beg, K.R., Al Mazeedi, H.M., Ali, L.N. and Saeed, T. (2006). Mycotoxins in poultry feed in Kuwait. *Archives of Environmental Contamination and Toxicology.* **50**: 594-602.

Belli, N., Ramos, A.J., Coronas, I., Sanchis, V. and Marin, S. (2005). *Aspergillus carbonarius* growth and ochratoxin A production on a synthetic grape medium in relation to environmental factors. *Journal of Applied Microbiology* **98**: 839-844.

Bennett, J.W. and Klich, M. (2003). Mycotoxins. *Clinical Microbiology Reviews* **16**: 497-516.

Bermudez, A.J., Ledoux, D.R. and Rottinghaus, G.E. (1995). Effects of *Fusarium moniliforme* culture material containing known levels of fumonisin B1 in duckling. *Avian Disease* **39**: 879-886.

Bermudez, A.J., Ledoux, D.R., Rottinghaus, G.E., Stosdill, P.L. and Bennett, G.A. (1997). Effects of feeding *Fusarium fujikuroi* culture material containing known levels of moniliformin in turkey poults. *Avian Pathology* **26**: 565-577.

Bersjo, B., Herstad, O. and Nafstad, I. (1993). Effects of feeding deoxynivalenol contaminated oats on reproduction performance in white leghorn hens. *British Poultry Science* **34**: 147-159.

Bezuidenhout, S.C., Gelderblom, W.C.A., Gorst-Allman, C.P., Horak, R.M., Marasas, W.F.O., Spiteller, G. and Vleggaar, R. (1988). Structure elucidation of the fumonisins, mycotoxins from *Fusarium moniliforme*. *Journal of the Chemical Society, Chemical Communications* **1988**: 743-745.

Bhatnagar, D., Yu, J. and Ehrlich, K.C. (2002). Toxins of filamentous fungi. *Chemical Immunology* **81**: 167-206.

Boonchuvit, B., Hamilton, P.B. and Burmeister, H.R. (1975). Interaction of T-2 toxin with *Salmonella* infections of chickens. *Poultry Science* **54**: 1693-1696.

Broomhead, J.N., Ledoux, D.R., Bermudez, A.J. and Rottinghaus, G.E. (2002). Chronic effects of fumonisin B1 in broilers and turkeys fed dietary treatments to market age. *Poultry Science* **81:** 56-61.

Brown, T.P., Rottinghaus, G.E. and Williams, M.E. (1992). Fumonisin mycotoxicosis in broiler: performance and pathology. *Avian Diseases* **36**: 450-454.

Burditt, S., Hagler, W.M. and Hamilton, P.B. (1984). Feed refusal during ochratoxicosis in turkeys. *Poultry Science* **63**: 2172-2174.

Burmeister, H.R., Ciegler, A. and Vesonder, R.F. (1979). Moniliformin, a metabolite of *Fusarium moniliforme* NRRL 6322: purification and toxicity. *Applied and Environmental Microbiology* **37**: 11-13.

Campbell, M.L. Jr., May, J.D., Huff, W.E. and Doerr, J.A. (1983). Evaluation of immunity of young broiler chickens during simultaneous aflatoxicosis and ochratoxicosis. *Poultry Science* **62**: 2138-2144.

Cespedes A.E. and Diaz, G.J. (1997). Analysis of aflatoxins in poultry and pig feeds and feedstuffs used in Colombia. *Journal of the AOAC International* **80**: 1215-1219.

Chang, C.F., Doerr, J.A. and Hamilton, P.B. (1981). Experimental ochratoxicosis in turkey poults. *Poultry Science* **60**: 114-119.

Cheng, Y.H., Shen, T.F., Pang, V.F. and Chen, B.J. (2001). Effects of aflatoxin and carotenoïde on growth performance and immune systeem in mule duikeling. *Comparative Biochemistry and Physiology Part C: Toxicology & Pharmacology* **128**: 19-26.

Chi, M.S., Mirocha, C.J., Kurtz, H.J., Weaver, G., Bates, F. and Shimoda, W. (1977). Sub-acute toxicity of T-2 toxin in broiler chicks. *Poultry Science* **56**: 306-313.

Chi, M.S., Robison, T.S., Mirocha, C.J. and Reddy, K.R. (1978). Acute toxicity of 12,13-epoxytrichothecenes in one day old broiler chicks. *Applied and Environmental Microbiology* **35**: 636-640.

Chowdhury, S.R., Smith, T.K., Boermans, H.J., Sefton, A.E., Downey, R. and Woodward, B. (2005a). Effects of feeding blends of grains naturally contaminated with *Fusarium* mycotoxins on performances, metabolism, haematology and immunocompetences of ducklings. *Poultry Science* **84**: 1179-1185.

Chowdhury, S.R., Smith, T.K., Boermans, H.J. and Woodward, B. (2005b). Effects of feedborne *Fusarium* mycotoxins on haematology and immunology of turkeys. *Poultry Science* **84**: 1698-1706.

Codex Alimentarius (2002). Proposed draft code of practice for the prevention of mycotoxin contamination in cereals. Joint FAO/WHO standards program, Rotterdam, The Netherlands.

Cole, R.J. (1986). Etiology of turkey X disease in retrospect: a case for the involvement of cyclopiazonic acid. *Mycotoxin Research* **2**: 3-7.

Conkova, E., Laciakova, A., Kovac, G. and Seidel, H. (2003). Fusarial toxins and their role in animal diseases. *Veterinary Journal* **165**: 214-220.

Conseil Superieur d'Hygiène Publique de France (CSHPF) (1999). *Les mycotoxines dans l'alimentation: évaluation et gestion du risqué.* éds Tec&Doc, Paris, France.

Corrier, D.E. (1991). Mycotoxicosis: mechanisms of immunosuppression. *Veterinary Immunology and Immunopathology* **30**: 73-87.

Cotty, P.J. and Jaime-Garcia, R. (2007). Influences of climate on aflatoxin producing fungi and aflatoxin contamination. *International Journal of Food Microbiology* **119**: 109-115.

Danicke, S., Valenta, H., Ueberschar, K.H. and Matthes, S. (2007). On the interaction between *Fusarium* toxin contaminated wheat and non-starch-polysaccharide hydrolysing enzymes in turkey diets on performance, health and carry over of deoxynivalenol and zearalenone. *British Poultry Science* **48**: 39-48.

Dalvi, R.R. (1986). An overview of aflatoxicosis of poultry: its characteristics, prevention and reduction. *Veterinary Research Communications* **10**: 429-443.

Diaz, G.J., Squires, E.J., Julian, R.J. and Boermans, H.J. (1994). Individual and combined effects of T-2 toxin and DAS in laying hens. *British Poultry Science* **35**: 393-405.

Doerr, J.A., Huff, W.E., Hamilton, P.B. and Lillehoj, E.B. (1981). Severe coagulopathy in young chickens produced by ochratoxin A. *Toxicology and Applied Pharmacology* **59**: 157-163.

Dombrink-Kurtzman, M.A., Javed, T., Bennett, G.A., Richard, LJ.L.,, Cote, L.M. and Buck, W.B. (1993). Lymphocyte cytotoxicity and erythrocytic abnormalities induced in broiler chicks by fumonisins B1 and B2 and moniliformin from Fusarium proliferatum. *Mycopathologia* **124**: 47-54.

Dorner, J.W., Cole, R.J., Lomax, L.G., Gosser, H.S. and Diener, U.L. (1983). Cyclopiazonic acid production by Aspergillus flavus and its effects on broiler chickens. *Applied and Environmental Microbiology* **46**: 698-703.

Duff, S.R., Burns, R.B. and Dwivedi, P. (1987). Skeletal changes in broiler chicks and turkey poults fed diets containing ochratoxin A. *Research in Veterinary Science* **43**: 301-307.

Dwivedi, P. and Burns, R.B. (1984a). Pathology of ochratoxicosis A in young broiler chickens. *Research in Veterinary Science* **36**: 92-103.

Dwivedi, P. and Burns, R.B. (1984b). Effect of ochratoxin A on immunoglobulins in broiler chicks. *Research in Veterinary Science* **36**: 117-121.

Edwards, S.G. (2004). Influence of agricultural practices on *Fusarium* infection of cereals and subsequent contamination of grain by trichothecene mycotoxins. *Toxicology Letters* **153**: 29-35.

Elissalde, M.H., Ziprin, R.L., Huff, W.E., Kubena, L.F. and Harvey, R.B. (1994). Effect of ochratoxin A on *Salmonella* challenged broiler chicks. *Poultry Science* **73**: 1241-1248.

Elling, F., Hald, B., Jacobsen, C. and Krogh, P. (1975). Spontaneous toxic nephropathy in country associated with ochratoxin A. *Acta Pathologica et Microbiologica Scandinavica Section A-Pathology* **83**: 739-741.

Espada, Y., Ruiz de Gopegui, R., Cuadradas, C. and Cabanes, F.J. (1994). Fumonisin mycotoxicosis in boilers. Weight and serum geriatrie modificatieons. *Avian Diseases* **38**: 454-460.

European Union Recommendation (2006). Prevention and reduction of Fusarium toxins in cereals and cereal-based products. *Official Journal of the European Union* **L234**: 35-40.

Fotso, J., Leslie, J.F. and Smith, J.S. (2002). Production of beauvericin, moniliformin, fusaroproliferin and fumonisins by fifteen ex-type strains of *Fusarium* species. *Applied and Environmental Microbiology* **68**: 5195-5197.

Fraga, M.E., Curvello, F., Gatti, M.J., Cavaglieri, L.R., Dalcero, A.M. and da Rocha Rosa, C.A. (2007). Potential aflatoxin and ochratoxin A production by *Aspergillus* species in poultry feed processing. *Veterinary Research Communications* **31**: 343-353.

Fukata, T., Sasai, K., Baba, E. and Arakawa, A. (1996). Effect of ochratoxin A on *Salmonella typhimurium* challenged layer chickens. *Avian Diseases* **40**: 924-926.

Galtier, P., Oswald, I.P., Guerre, P., Morgavi, D., Boudra, H. and Jouany, J.P. (2008). Le risque mycotoxique: danger et impact sanitaire en productions animales. *INRA Productions Animales* **21**: 107-116.

Gareis, M., Schothorst, R., Vidnes, A., Bergsten, C., Paulsen, B., Brera, C. and Miraglia, M. (2003). Collection of occurrence data of *Fusarium* toxins in food. Available at: www.europa.eu.int/comm/food/fs/scoop/3.2.8_en.pdf

Gaumy, J.L., Billy, J.D., Benrad, G. and Guerre, P. (2001). Zearalenone: origine et effets chez les animaux. *Revue de Médecine Vétérinaire* **152**: 123-136.

Gelderblom, W.C., Jaskiewicz, K., Marasas, W.F., Thiel, P.G., Horak, R.M., Vleggaar, R. and Kriek, N.P. (1988). Fumonisin-novel mycotoxins with cancer-promoting activity produced by *Fusarium moniliforme*. *Applied and Environmental Microbiology* **54**: 1806-1811.

Gentles A., Smith, E.E., Kubena, L.F., Duffus, E., Johnson, P., Thompson, J., Harvey, R.B. and Edrington, T.S. (1999). Toxicological evaluations of cyclopiazonic acid and ochratoxin A in broilers. *Poultry Science* **78**: 1380-1384.

Ghosh, R.C., Chauhan, H.V. and Jha, G.J. (1991). Suppression of cell-mediated immunity by purified aflatoxin B1 in broiler chicks. *Veterinary Immunology and Immunopathology* **28**: 165-172

Giambrone, J.J., Diener, U.L., Davis, N.D., Panangala, V.S. and Hoerr, F.J. (1985). Effects of purified aflatoxin on broiler chickens. *Poultry Science* **64**: 852-858.

Gibson, R.M., Bailey, C.A., Kubena, L.F., Huff, W.E. and Harvey, R.B. (1989). Ochratoxin A and dietary protein. Effect on body weight, feed conversion, relative organ weight and mortality in three old broilers. *Poultry Science* **68**: 1658-1663.

Gibson, R.M., Bailey, C.A., Kubena, L.F., Huff, W.E. and Harvey, R.B. (1990). Impact of L-phenylalanine supplementation on the performance of three week old broilers fed diets containing ochratoxin A. Effect on body weight, feed conversion, relative organ weight, and mortality. *Poultry Science* **69**: 414-419.

Giorni, P., Magan, N., Pietri, A., Bertuzzi, T. and Battilani, P. (2007). Studies on *Aspergillus* section flavi isolated from maize in northern Italy. *International Journal of Food Microbiology* **113**: 330-338.

Girish, C.K., Macdonald, E.J., Scheinin, M. and Smith, T.K. (2008a). Effects of feedborne *Fusarium* mycotoxins on brain regional neurochemistry of turkeys. *Poultry Science* **87**: 1295-1302.

Girish, C.K. and Smith, T.K. (2008). Effects of feeding blends of grains naturally contaminated with *Fusarium* mycotoxins on small intestinal morphology of turkeys. *Poultry Science* **87**: 1075-1082.

Girish, C.K., Smith, T.K., Boermans, H.J. and Karrow, N.A. (2008b). Effects of feeding blends of grains naturally contaminated with *Fusarium* mycotoxins on performance, haematology, metabolism and immunocompetence of turkeys. *Poultry Science* **87**: 421-432.

Glahn, R.P., Wideman, R.F. jr, Evangelisti, J.W. and Huff, W.E. (1988). Effects of ochratoxin A alone and in combination with citrinin on kidney function of single comb white leghorn pullets. *Poultry Science* **67**: 1034-1042.

Guerre, P. (2000). Intérêt des traitements des matières premières et de l'usage d'adsorbants lors d'une contamination des aliments du bétail par des mycotoxines. *Revue de Médecine Vétérinaire* **151**: 1095-1106.

Hamilton, P.B. (1971). A natural and extremely severe occurrence of aflatoxicosis in laying hens. *Poultry Science* **50**: 1880-1882.

Hamilton, P.B. (1975). Proof of mycotoxicoses being a filed problem and a simple method for their control. *Poultry Science* **54**: 1206-1208.

Hamilton, P.B., Huff, W.E., Harris, J.R. and Wyatt, R.D. (1982). Natural occurrences of ochratoxicosis in poultry. *Poultry Science* **61**: 1832-1841.

Hamilton, R.M.G., Thompson, B.K., Trenholm, H.L., Fiser, P.S. and Greenhalgh, R. (1985). Effects of feeding white leghorn hens diets that contain deoxynivalenol-contaminated wheat. *Poultry Science* **64**: 1840-1852.

Harvey, R.B., Kubena, L.B., Rottinghaus, G.E., Turk, J.R., Casper, H.H. and Buckley, S.A. (1997). Moniliformin from *Fusarium fujikuroi* culture material and deoxynivalenol from naturally contaminated wheat incorporated into diets of broiler chicks. *Avian Diseases* **41**: 957-963.

Henry, M.H., Wyatt, R.D. and Fletchert, O.J. (2000). The toxicity of purified fumonisin B1 in broiler chicks. *Poultry Science* **79**: 1378-1384.

Hollinger, K. and Ekperigin, H.E. (1999). Mycotoxicosis in food producing animals. *Veterinary Clinics of North America: Food Animal Practice* **15**: 133-165.

Huff, W.E. and Ruff, M.D. (1982). *Eimeria acervulina* and *Eimeria tenella* infectieons in ochratoxin A compromised broiler chickens. *Poultry Science* **61**: 685-692.

Huff, W.E., Doerr, J.A., Wabeck, C.J., Chalouka, G.W., May, J.D. and Merkley, J.W. (1984). The individual and combined effects of aflatoxin and ochratoxin A on various processing parameters of broiler chickens. *Poultry Science* **63**: 2153-2161.

Huff, W.E., Harvey, R.B., Kubena, L.F. and Rottinghaus, G.E. (1988a). Toxic synergism between aflatoxin and T-2 toxin in broiler chickens. *Poultry Science* **67**: 1418-1423.

Huff, W.E., Kubena, L.F. and Harvey, B. (1988b). Progression of ochratoxicosis in broiler chickens. *Poultry Science* **67**: 1139-1146.

Huff, W.E., Kubena, L.F., Harvey, R.B., Hagler, W.M. jr., Swanson, S.P., Phillips, T.D. and Creger, C.R. (1986). Individual and combined effects of aflatoxin and deoxynivalenol in broiler chickens. *Poultry Science* **65**: 1291-1298.

Huff, W.E., Wyatt, R.D. and Hamilton, P.B. (1975). Nephrotoxicity of dietary ochratoxin A in broiler chickens. *Applied Microbiology* **30**: 48-51.

Huff, W.E., Wyatt, R.D., Tucker, T.L. and Hamilton, P.B. (1974). Ochratoxicosis in the broiler chicken. *Poultry Science* **53**: 1585-1591.

Hulan, H.W. and Proudfoot, F.G. (1982). Effects of feeding vomitoxin contaminated wheat on performance of broiler chickens. *Poultry Science* **61**: 1653-1659.

Hussein, H.M., Baxter, M., Andrew, I.G. and Franich, R.A. (1991). Mycotoxin production by *Fusarium* species isolated from New Zealand maize fields. *Mycopathologia* **113**: 35-40.

Huwig, A., Freimund, S., Kappeli, O. and Dutler, H. (2001). Mycotoxin detoxication of animal feed by different adsorbents. *Toxicology Letters* **122**: 179-188.

IARC (1993). Some naturally occurring substances, food items and constituents, heterocyclic aromatic amines and mycotoxins. In: *Monographs on the evaluation of carcinogenic risks to humans 56*, World health organization, Lyon, p. 245.

Javed, T., Dombrink-Kurtzman, M.A., Richard, J.L., Bennett, G.A., Cote, L.M. and Buck, W.B. (1995). Serohamatologic alterations in broiler chicks on feed amended with *Fusarium proliferatum* culture material on fumonisin B1 and moniliformin. *Journal of Veterinary Diagnostic Investigation* **7**: 520-526.

Jorgensen, K. (2005). Occurrence of ochratoxin A in commodities and processed food: a review of EU occurrence data. *Food Additives and Contaminants* **22**: 26-30.

Kaaya, A.N. and Kyamuhangire, W. (2006). The effect of storage time and agroecological zone on mould incidence and aflatoxin contamination of maize from traders in Uganda. *International Journal of Food Microbiology*, **110**: 217-223.

Kabak, B., Dobson, A.D.W. and Var, I. (2006). Strategies to prevent mycotoxin contamination of food and animal feed: a review. *Critical Reviews in Food Science and Nutrition* **46**: 593-619.

Karaman, M., Basmacioglu, H., Ortatatli, M. and Oguz, H. (2005). Evaluation of the detoxifying effect of yeast glucomannan on alfatoxicosis in boilers as assessed by gross examinator and histopathology. *British Poultry Science* **46**: 394-400.

Klein, P.J., Van Vleet, T.R., Hall, J.O. and Coulombe, R.A. Jr. (2002). Biochemical factors underlying the age-related sensitivity of turkeys to aflatoxin B1. *Comparative Biochemistry and Physiology Part C: Toxicology & Pharmacology* **132**: 193-201.

Klich, M.A. and Pitt, J.I. (1968). Differentiation of *Aspergillus flavus* from *Aspergillus parasiticus* and other closely related species. *Transactions of the British Mycological Society* **91**: 99-108.

Kolf-Clauw, M., Avouni, F., Tardieu D. and Guerre P. (2008). Variation in zearalenone activation in avian food species. *Food and Chemical Toxicology* **46**: 1467-1473.

Kozaczynski, W. (1994). Experimental ochratoxicosis A in chicken. Histopathological and histochemical study. *Archivum Veterinarium Polonicum* **34**: 205-219.

Kubena, L.F., Harvey, R.B., Fletcher, O.J., Phillips, T.D., Mollenhauer, H.H., Witzel, D.A. and Heidelbaugh, N.D. (1985). Toxicity of ochratoxin A and vanadium to growing chicks. *Poultry Science* **64**: 620-628.

Kubena, L.F., Edrington, T.S., Harvey, R.B., Buckley, S.A., Phillips, T.D., Rottinghaus, G.E. and Casper, H.H. (1997). Individual and combined effects of fumonisin B1 present in *Fusarium moniliforme* culture material and deoxynivalenol in broiler chicks. *Poultry Science* **76**: 1239-1247.

Kubena, L.F., Edrington, T.S., Harvey, R.B., Phillips, T.D., Sarr, A.B. and Rottinghaus, G.E. (1997). Individual and combined effects of fumonisin B1 present in *Fusarium moniliforme* culture material and diacetoxyscirpenol or ochratoxin A in turkey poults. *Poultry Science* **76**: 256-264.

Kubena, L.F., Edrington, T.S., Kamps-Holtzapple, C., Harvey, R.B., Elissalde, M.H. and Rottinghaus, G.E. (1995). Effect of feeding fumonisin B1 present in *Fusarium moniliforme* culture material and aflatoxin singly and in combination to turkey poults. *Poultry Science* **74**: 1295-1303.

Kubena, L.F., Harvey, R.B., Buckley, S.A., Bailey, R.H. and Rottinghaus, G.E. (1999). Effects of long term feeding of diets containing moniliformin, supplied by *Fusarium fujokuroi* culture material and fumonisin supplied by *Fusarium moniliforme* culture material to laying hens. *Poultry Science* **78**: 1499-1505.

Kubena, L.F., Harvey, R.B., Edrington, T.S. and Rottinghaus, G.E. (1994). Influence of ochratoxin A and diacetoxyscirpenol singly and in combination on broiler chickens. *Poultry Science* **73**: 408-415.

Kubena, L.F., Harvey, R.B., Huff, W.E., Corrier, D.E., Phillips, T.D. and Rottinghaus, G.E. (1989). Influence of ochratoxin A and T-2 toxin singly and in combination on broiler chickens. *Poultry Science* **68**: 867-872.

Kubena, L.F., Huff, W.E., Harvey, R.B., Phillips, T.D. and Rottinghaus, G.E. (1989). Individual and combined toxicity of deoxynivalenol and T-2 toxin in broiler chicks. *Poultry Science* **68**: 622-626.

Kumar, A., Jindal, N., Shukla, C.L., Asrani, R.K., Ledoux, D.R. and Rottinghaus, G.E. (2004). Pathological changes in broiler chickens fed ochratoxin A and inoculated with *Escherichia coli*. *Avian Pathology* **33**: 413-417.

Kumar, A., Jindal, N., Shukla, C.L., Pal, Y., Ledoux, D.R. and Rottinghaus, G.E. (2003). Effect of ochratoxin A on *Escherichia coli* challenged broiler chicks. *Avian Diseases* **47**: 415-424.

Labuda, R., Parich, A., Berthiller, F. and Tancinova, D. (2005). Incidence of trichothecenes and zearalenone in poultry feed mixtures from Slovakia. *International Journal of Food Microbiology* **105**: 19-25.

Labuda, R., Parich, A., Vekiru, E. and Tancinova, D. (2005). Incidence of fumonisins, moniliformin and *Fusarium* species in poultry feed mixtures from Slovakia. *Annals of Agricultural and Environmental Medicine* **12**: 81-86.

Le Bars, J. (1979). Cyclopiazonic acid production by *Penicillium camemberti* Thom and natural occurrence of this mycotoxin in cheese. *Applied and Environmental Microbiology* **38**: 1052-1055.

Le Bars, J., Le Bars P., Dupuy, J., Boudra H. and Cassini, R. (1994). Biotic and abiotic factors in fumonisin B1 production and stability. *Journal of the AOAC International* **77**: 517-521.

Leblanc, J.C., Tard, A., Volatier, J.L. and Verger, P. (2005). Estimated dietary exposure to principal food mycotoxins from the first French total diet study. *Food Additives and Contaminants* **22**: 652-672.

Ledoux, D.R., Bermudez, A.J. and Rottinghaus, G.E. (1996). Effects of feeding *Fusarium moniliforme* culture material containing known levels of fumonisin B1 in the young turkey poult. *Poultry Science* **75**: 1472-1478.

Ledoux, D.R., Broomhead, J.N., Bermudez, A.J. and Rottinghaus, G.E. (2003). Individual and combined effects of the *Fusarium* mycotoxins fumonisin B1 and moniliformin in broiler chicks. *Avian Diseases* **47**: 1368-1375.

Ledoux, D.R., Brown, T.P., Weibking, T.S. and Rottinghaus, G.E. (1992). Fumonisin toxicity in broiler chicks. *Journal of Veterinary Diagnostic Investigation* **4**: 330-3.

Li, Y.C., Ledoux, D.R., Bermudez, A.J., Fritsche, K.L. and Rottinghaus, G.E. (2000). The individual and combined effect of fumonisin B1 and moniliformin on performance and selected immune parameters in turkey poults. *Poultry Science* **79**: 871-878.

Li, Y.C., Ledoux, D.R., Bermudez, A.J., Fritsche, K.L. and Rottinghaus, G.E. (2000). The individual and combined effects of moniliformin on performance and immune function of broiler chicks. *Poultry Science* **79**: 26-32.

Llorens, A., Mateo, R., Hinojo, M.J., Logrieco, A. and Jimenez, M. (2004). Influence of the interactions among ecological variables in the characterization of zearalenone producing isolates of *Fusarium* spp. *Systematic and Applied Microbiology* **27**: 253-260.

Logrieco, A., Mule, G., Moretti, A. and Bottalico, A. (2002). Toxigenic *Fusarium* species and mycotoxins associated with maize ear rot in Europe. *European Journal of Plant Pathology* **108**: 597-609.

Lozano, M.C. and Diaz, G.J. (2006). Microsomal and cytosolic biotransformation of aflatoxin B1 in four poultry species. *British Poultry Science* **47**: 734-741.

Magan, N. and Aldred, D. (2005). Conditions of formation of ochratoxin A in drying, transport, and in different commodities. *Food Additives and Contaminants* **22**: 10-16.

Magan, N., Hope, R., Cairns, V. and Aldred, D. (2003). Post-harvest fungal ecology: impact of fungal growth and mycotoxin accumulation in stored grain. *European Journal of Plant Pathology* **109**: 723-730.

Marin, S., Magan, N., Ramos, A.J. and Sanchis, V. (2004). Fumonisin-producing strains of *Fusarium*: a review of their ecophysiology. *Journal of Food Protection* **67**: 1792-1805.

Martins, M.L. and Martins, H.M. (1999). Natural and in vitro coproduction of cyclopiazonic acid and aflatoxins. *Journal of Food Protection* **62**: 292-294.

Meissonier, G.M., Marin, D.E., Galtier, P., Bertin, G., Taranu, I. and Oswald, I.P. (2006). Modulation of the immune response by a group of fungal food contaminant, the aflatoxins. In: *Nutrition and Immunity* (Eds. Mengheri, E., Roselli, M., Britti, M.S. and Finamore, A.). Research Signpost, Kerala, India, pp 147-166.

Miller, J.D. (2002). Aspects of the ecology of Fusarium toxins in cereals. *Advances in Experimental Medicine and Biology.* **504**: 19-27.

Molto, G.A., Gonzalez, H.H., Resnik, S.L. and Pereyra-Gonzalez, A. (1997). Production of trichothecenes and zearalenone by isolates of *Fusarium* spp. from Argentinian maize. *Food Additives and Contaminants* **14**: 263-268.

Nesbitt, B.F., O'Kelly, J., Sargeant, K. and Sheridan, A. (1962). *Aspergillus flavus* and turkey X disease. Toxic metabolites of *Aspergillus flavus*. *Nature* **195**: 1062-1063.

Newberne, P.M. and Butler, W.H. (1969). Acute and chronic effects of aflatoxin on the liver of domestic and laboratory animals: a review. *Cancer Research* **29**: 236-250.

Nizamlyoglu, F. and Oguz, H. (2003). Occurrence of aflatoxins in layer feed and corn samples in Konya province, Turkey. *Food Additives and Contaminants* **20**: 654-658.

Obst, A., Lepschy, J., Beck, R., Bauer, G. and Bechtel, A. (2000). The risk of toxins by *Fusarium graminearum* in wheat – interactions between weather and agronomic factors. *Mycotoxin Research* **16**: 16-20.

Oliveira, G.R., Ribeiro, J.M., Fraga, M.E., Cavaglier, L.R., Direito, G.M., Keller, K.M., Dalcero, A.M. and Rosa, C.A. (2006). Mycobiota in poultry feeds and natural occurrence of aflatoxins, fumonisins and zearalenone in the Rio de Janeiro state, Brazil. *Mycopathologia* **162**: 355-362.

Ortatatli, M. and Oguz, H. (2001). Ameliorative effect of dietary clinoptilolite on pathological changes in broiler chickens during aflatoxicosis. *Research in Veterinary Science* **71**: 59-66.

Page, R.K., Stewart, G., Wyatt, R., Bush, P., Fletcher, O.J. and Brown, J. (1980). Influence of low levels of ochratoxin A on egg production, egg shell stains and serum uric-acid in leghorn type hens. *Avian Diseases* **24**: 777-780.

Pardo, E., Marin, S., Ramos, A.J. and Sanchis, V. (2006). Ecophysiology of ochratoxigenic *Aspergillus ochraceus* and *Penicillium verrucosum* isolates; Predictive models for fungal spoilage prevention: a review. *Food Additives and Contaminants* **23**: 398-410.

Park, D.L. (2002). Effect of processing on aflatoxin. *Advances in Experimental Medicine and Biology* **54**: 173-179.

Parry, D.W., Jenkinson, P. and McLeod, L. (1995). *Fusarium* ear blight (scab) is small grain cereals-a review. *Plant Pathology* **44**: 207-238.

Pitt, J.I. (2002). Biology and ecology of toxigenic *Penicillium* species. *Advances in Experimental Medicine and Biology* **504**: 29-41.

Pitt, J.I. (1987). *Penicillium viridicatum, Penicillium verrucosum*, and production of ochratoxin A. *Applied and Environmental Microbiology* **53**: 266-269.

Piva, G., Galvano, F., Pietri, A. and Piva, A. (1995). Detoxification methods of aflatoxins: a review. *Nutrition Research* **15**: 767-776.

Prathapkumar, S.H., Rao, V.S., Paramkishan, R.J. and Bhat, R.V. (1997). Disease outbreak in laying hens arising from the consumption of fumonisin-contaminated food. *British Poultry Science* **38**: 475-479.

Prior, M.G. and Sisodia, C.S. (1978). Ochratoxicosis in white Leghorn hens. *Poultry Science* **57**: 619-623.

Proctor, A.D., Ahmedna, M., Kumar, J.V. and Goktepe, I. (2004). Degradation of aflatoxins in peanut kernels/flour by gaseous ozonation and mild heat treatment. *Food Additives and Contaminants* **21**: 786-793.

Queshi, M.A., Brake, J., Hamilton P.B., Hagler, W.M.Jr. and Nesheim, S. (1998). Dietary exposure of broiler breeders to aflatoxin results in immune dysfunction in progeny chicks. *Poultry Science* **77**: 812-819.

Qureshi, M.A. and Hagler, W.M.Jr. (1992). Effect of fumonisin B1 exposure on chicken macrophage functions in vitro. *Poultry Science* **71**: 104-112.

Qureshi, M.A., Garlich, J.D., Hagler, W.M.Jr. and Weinstock, D. (1995). *Fusarium proliferatum* culture material alters several production and immune performance parameters in White Leghorn chickens. *Immunopharmacology and Immunotoxicology* **17**: 791-804.

Rafai, P., Pettersson, H., Bata, A., Papp, Z., Glavits, R., Tuboly, S., Vanyi, A. and Soos, P. (2000). Effect of dietary T-2 fusariotoxin concentrations on the health and production of white Pekin duck broilers. *Poultry Science* **79**: 1548-1556.

Rapper, K.B. and Fennel, D.I. (1965). *The genus Aspergillus.* Williams & Wilkins, Baltimore, Maryland.

Raynal, M., Bailly, J.D., Benard, G. and Guerre P. (2001). Effects of fumonisin B1 present in *Fusarium moniliforme* culture material on drug metabolising enzyme activities in ducks. *Toxicology Letters* **121:** 179-190.

Richard, J.L., Cysewwski, S.J., Pier, A.C. and Booth, G.D. (1978). Comparison of effects of dietary T-2 toxin on growth, immunogenic organs antibody formation and pathologic changes in turkeys and chickens. *American Journal of Veterinary Research* **39**: 1674-1679.

Richardson, K.E. and Hamilton, P.B. (1990). Comparative toxicity of scirpentriol and its acetylated derivatives. *Poultry Science* **69**: 397-402.

Schaeffer, J.L., Tyczkowski, J.K. and Hamilton, P.B. (1987). Alteration in caretenoid metabolism during ochratoxicosis in young broiler chickens. *Poultry Science* **66**: 318-324.

Schrödter, R. (2004). Influence of harvest and storage conditions on trichothecenes levels in various cereals. *Toxicology Letters* **153**: 47-49.

Scott, P.M. (1998). Industrial and farm detoxification processes for mycotoxins. *Revue de Médecine Vétérinaire* **149:** 543-548.

Sharbi, T.F., Templeton, G.E., Beasley J.N., Stephenson, E.L. (1973). Toxicity resulting from feeding experimentally molded corn to broiler chicks. *Poultry Science* **52**: 1007-1014.

Shivachandra, S.B., Sah, R.L., Singh, S.D., Kataria, J.M. and Manimaran, K. (2003). Immunosuppression in broiler chicks fed aflatoxin and inoculated with fowl adenovirus serotype 4 associated with hydropericardum syndrome. *Veterinary Research Communications* **27**: 39-51.

Sklan, D., Shelly, M., Makovsky, B., Geyra, A., Klipper, E. and Friedman, A. (2003). The effect of chronic feeding of diacetoxyscirpenol and T-2 toxin on performance, health, small intestinal physiology and antibody production in turkey poults. *British Poultry Science* **44**: 46-52.

Smith, E.E., Kubena, L.F., Braithwaite, C.E., Harvey, R.B., Phillips, T.D. and Reine, A.H. (1992). Toxicological evaluation of aflatoxin and cyclopiazonic acid in broiler chickens. *Poultry Science* **71**: 1136-1144.

Smith, J.W., Prince, W.R. and Hamilton,, P.B. (1986). Relationship of aflatoxicosis to Salmonella gallinarum infection of chickens. *Applied Microbiology* **18**: 946-947.

Stob, M., Baldwin, R.S., Tuite, J., Andrews, F.N. and Gillette, K.G. (1962). Isolation of an anabolic, uterotrophic compound from corn infected with *Gibberella zeae*. *Nature* **29**: 196.

Sydenham, E.W., Marasas, W.F., Thiel, P.G., Shephard, G.S. and Nieuwenhuis, J.J. (1991). Production of mycotoxins by selected *Fusarium graminearum* and *F. crookwellense* isolates. *Food Additives and Contaminants* **8**: 31-41.

Taniwaki, M.H. (2006). An update on ochratoxigenic fungi and ochratoxin A in coffee. *Advances in Experimental Medicine and Biology*. **571**: 189-202.

Tardieu, D., Auby, A., Bluteau C., Bailly J.D. and Guerre P. (2008). Determination of fumonisin B1 in animal tissues with immunoaffinity purification. *Journal of Chromatography B, Analytical Technologies in the Biomedical and Life Sciences* **870**: 140-144.

Tardieu, D., Bailly, J.D., Bénard, G., Tran, T.S. and Guerre, P. (2004). Toxicity of maize containing known levels of fumonisin B1 during force feeding of duck. *Poultry Science* **83**: 1287-1293.

Tardieu, D., Bailly, J.D., Skiba, F., Métayer, J.P., Grosjean, F. and Guerre P. (2007). Chronic toxicity of fumonisins in turkeys. *Poultry Science* **86**: 1887-1893.

Tardieu, D., Tran, S.T., Auvergne, A., Babilé, R., Bénard, G., Bailly, J.D. and Guerre P. (2006). Effects of fumonisins on liver and kidney sphinganine and the sphinganine to sphingosine ratio during chronic exposure in ducks. *Chemico-biological Interactions* **160**: 51-60.

Teren, J., Varga, J., Hamari, Z., Rinyu, E. and Kevei, F. (1996). Immunochemical detection of ochratoxin A in black *Aspergillus* strains. *Mycopathologia* **134**: 171-176.

Tessari, E.N., Oliveira, C.A., Cardoso, A.L., Ledoux, D.R. and Rottinghaus, G.E. (2006). Effects of aflatoxin B1 and fumonisin B1 on body weight, antibody titres and histology of broiler chicks. *British Poultry Science* **47**: 357-364.

Thompson, C. and Henke, S. (2000). Effect of climate and type of storage container on aflatoxin production in corn and its associated risks to wildlife species. *Journal of Wildlife Diseases* **36**: 172-179.

Tran S.T., Auvergne A., Benard G., Bailly, J.D., Tardieu, D., Babilé R. and Guerre, P. (2005). Chronic effect of fumonisin B1 on ducks. *Poultry Science* **84**: 22-28.

Tran, S.T., Bailly, J.D., Tardieu, D., Durand, S., Benard, G. and Guerre, P. (2003). Sphinganine to sphingosine ratio as predictive biochemical markers of fumonisin B1 exposure in ducks. *Chemico-biological Interactions* **146**: 61-72.

Tran, S.T., Tardieu, D., Auvergne, A., Bailly, J.D., Babilé, R., Durand, S., Benard, G. and Guerre, P. (2006). Serum sphinganine and the sphinganine to sphingosine ratio during chronic exposure in ducks. *Chemico-biological Interactions* **160**: 41-50.

Van der Merwe, K.J., Steyn P.S., Fourie, L., Scott D.B. and Theron, J.J. (1965). Ochratoxin A, a toxic metabolite produced by *Aspergillus ochraceus* Wilh. *Nature* **205**: 1112-1113.

Vanyi, A., Bata, A. and Kovacs, F. (1994). Effects of T-2 toxin treatment on the egg yield and hatchability in geese. *Acta Veterinaria Hungarica* **42**: 79-85.

Verma, J., Johri, T.S., Swain, B.K. and Ameena, S. (2004). Effect of graded levels of aflatoxin, ochratoxin and their combinations on the performance and immune response of broilers. *British Poultry Science* **45**: 512-518.

Vesely, D., Vesela D. and Neumannova, V. (1985). Production of cyclopiazonic acid, its effect on the chock embryo and on 1-day old cockerels. *Veterinární Medicína (Praha)* **30**: 345-352.

Virdi, J.S., Tiwari, R.P., Saxena, M., Khanna, V., Singh G., Saini, S.S. and Vadehra, D.V. (1989). Effects of aflatoxin on the immune system of the chick. *Journal of Applied Toxicology* **9**: 271-275.

Warren, M.F. and Hamilton, P.B. (1980). Intestinal fragility during ochratoxicosis and aflatoxicosis in broiler chickens. *Applied and Environmental Microbiology* **40**: 641-645.

Weibking, T.S., Ledoux, D.R., Bermudez, A.J., Turk, J.R., Rottinghaus, G.E., Wang, E. and Merrill, A.H.,Jr. (1993). Effects of feeding *Fusarium moniliforme* culture material containing known levels of fumonisin B1 on the young broiler chick. *Poultry Science* **72**: 456-466.

Wilkins, K., Nielsen, K.F. and Din, S.U. (2003). Patterns of volatile metabolites and non-volatile trichothecenes produced by isolates of *Stachybotrys*, *Fusarium*, *Trichoderma*, *Trichothecium* and *Memnoniella*. *Environmental Science and Pollution Research International* **10**: 162-166.

Wilson, D.M., Mubatanhema, W. and Jurjevic, Z. (2002). Biology and ecology of mycotoxigenic Aspergillus species as related to economic and health concerns. *Advances in Experimental Medicine and Biology* **504**: 3-17.

Wu, F. (2006). Mycotoxin reduction in Bt corn: potential economic, health and regulatory impacts. *Transgenic Research* **15**: 277-289.

Wyatt, R.D. and Hamilton, P.B. (1975). Interaction between aflatoxicosis and a natural infection of chickens with Salmonella. *Applied Microbiology* **30**: 870-872.

Wyatt, R.D., Hamilton, P.B. and Burmeister, H.R. (1973). The effect of T-2 toxin in broiler chickens. *Poultry Science* **52**: 1853-1859.

Yegani, M., Smith, T.K., Leeson, S., and Boermans, H.J. (2006). Effects of feeding grains naturally contaminated with *Fusarium* mycotoxins on performances and metabolism of broiler breeders. *Poultry Science* **85**: 1541-1549.

Young, J.C., Zhu, H. and Zhou, T. (2006). Degradation of mycotoxins by aqueous ozone. *Food and Chemical Toxicology* **44**: 417-424.

Zinedine, A., Juan, C., Soriano, J.M., Molto, J.C., Idrissi, L. and Manes, J. (2007). Limited survey for the occurrence of aflatoxins in cereals and poultry feeds from Rabat, Morocco. *International Journal of Food Microbiology* **115**: 124-127.

Keyword index

A

acidic compounds 202
aflatoxin 79, 232
   – B$_1$ 90
   – breeding performance 233
   – chronic toxicity 233
   – immunosuppression 234
AI *See:* avian influenza
alkaline phosphatase 14
all-in/all-out
   – 108
amino acids
   – deficiency 63
   – release 151
   – sulfur 36
ammonia 54
ANFs *See:* anti-nutritive factors
animal
   – health 231
   – performance 139
anti-nutritive factors (ANFs) 32
antibiotic-resistance plasmids 129
antibiotic growth promoters 168
antibiotics 18
   – alternatives 52
antigen 15
antimicrobial growth promoters 122
antimicrobials 215, 219
arginine 37, 57, 60, 149
   – deficiency 39
avian influenza (AI) 221

B

bacteria
   – gram-positive 16
   – prophylactic properties 19
   – therapeutic properties 19
bacterial adhesion 125

barley 52
best management practices 226
Bio-Mos® 126, 200
biosecurity 214, 218, 221
   – cost effectiveness 222
black head 218
bursa 140, 145, 147
butyrate 54, 202

C

carbohydrates 124, 199
carboxymethylcellulose (CMC) 53
CAT *See:* cationic amino acid
   transporters
cationic amino acid transporters (CAT)
   145
cells
   – epithelial 14
   – goblet 13, 32, 62
   – mucosal dendritic 15
   – secretory 13
cereals, contamination 249
chitin 87
Clostridium perfringens 168, 216
CMC *See:* carboxymethylcellulose
coccidiosis 172, 176
Cochlosoma anatis 172
commensal microbiota 13, 16
competitive exclusion 197, 201
complex-forming capacity 83
crop rotation 251
crypt depth 32
cyclopiazonic acid 246
cysteine 35, 150

D

DAS *See:* diacetoxyscripenol
deoxynivalenol (DON) 91, 241